State of The Space Industrial Base 2021

Infrastructure & Services for Economic Growth & National Security

J. Olson, S. Butow, E. Felt,
T. Cooley, & J. Mozer

Nimble Books LLC: The AI Lab for Book-Lovers
🚀 Fred Zimmerman, Editor 🚀

Humans and AI making books richer, more diverse, and more surprising.

Publishing Information

(c) 2023 Nimble Books LLC
ISBN: 978-1-60888-242-7

AI-generated Keyword Phrases

space industrial base; workshop report; state of the space industry; United States; national vision and strategy; diversity in space industry; national power; competitiveness in space domain; key issues and challenges; future actions; current state of space industrial base; improvement recommendations; technology convergence; government-private sector partnerships; innovation and growth in space industry; long-term national space strategy; commercial space activities.

Front Matter

- Publishing Information ii
- AI-generated Keyword Phrases ii
- Abstracts iv
  - TL;DR (one word) iv
  - Explain It To Me Like I'm Five Years Old iv
  - TL;DR (vanilla) iv
  - Scientific Style iv
  - Action Items v
- Viewpoints vi
  - Grounds for Dissent vi
  - Red Team Critique vii
  - MAGA Perspective viii
- Summaries x

ABSTRACTS

TL;DR (ONE WORD)

Space industry.

EXPLAIN IT TO ME LIKE I'M FIVE YEARS OLD

This document is like a report card about the space industry. It tells us how things are going and gives suggestions for the future. The report says that it's important for the United States to have a clear plan and be a leader in space. It also says that it's important for everyone in America and our friends to be involved in space activities. The report talks about how space is really powerful and can make a country strong. It also talks about the challenges the space industry faces, like competition,

TL;DR (VANILLA)

This report summarizes the state of the space industry in 2021 and provides recommendations for its improvement. It emphasizes the need for a clear national vision and strategy, diversity and inclusion, competitiveness, partnerships between government and private sector, and a long-term national space strategy. The report also highlights the potential of space as a major factor in national power and predicts that commercial space activities may surpass government spending in the future.

SCIENTIFIC STYLE

This summary report provides insights and recommendations on the state and future of the space industry, with a focus on the United States. The report emphasizes the need for a clear national vision and strategy for space development and leadership, as well as the importance of diversity and collaboration in this endeavor. It discusses key challenges and issues facing the industry, and proposes recommendations for improvement. The document highlights the convergence of technology and the importance of government-private sector partnerships in driving innovation and growth in the space industry. It also underscores the need for a long-term national space strategy and recognizes the potential for commercial space activities to surpass government spending in the future.

ACTION ITEMS

Develop a clear national vision and strategy for space development and leadership, involving input from various stakeholders including industry experts, government officials, and academia.

Foster partnerships between the government and the private sector to drive innovation and growth in the space industry.

Invest in research and development to maintain competitiveness in the face of increased global competition.

Promote diversity and inclusion in the space industry to ensure a wide range of perspectives and talent.

Develop a long-term national space strategy that outlines goals, objectives, and milestones for space development and exploration.

Support and incentivize commercial space activities to stimulate economic growth.

VIEWPOINTS

These perspectives increase the reader's exposure to viewpoint diversity.

GROUNDS FOR DISSENT

A member of the organization responsible for this document might have principled, substantive reasons to dissent from this report for several potential reasons:

Lack of emphasis on international collaboration: The report highlights the importance of engaging America and its allies in space development but may not sufficiently emphasize the significance of international collaboration. A dissenting member may argue that global cooperation is crucial for the advancement of space exploration and innovation and that the report should place greater emphasis on fostering partnerships with other nations, particularly those that have expertise in specific areas.

Overemphasis on national power: The document mentions the potential of space as a major factor in national power. However, a dissenting member may hold the view that prioritizing national power could lead to increased militarization of space or create an environment of competition rather than cooperation. They may argue for a more balanced approach that values international collaboration and focuses on using space as a tool for global welfare and scientific advancement.

Insufficient consideration of ethical implications: The report primarily focuses on the economic and technological aspects of space development, but a dissenting member may believe that it fails to adequately address the ethical considerations associated with space exploration. They may argue for a stronger emphasis on responsible space practices, such as minimizing space debris, considering the impact on celestial bodies, and ensuring equitable access to space resources.

Neglect of sustainability and environmental impacts: The report may not sufficiently address the potential environmental impacts of space activities. A dissenting member may argue for stronger recommendations regarding sustainable practices in space exploration, including mitigating the effects of rocket launches, managing space debris, and minimizing the use of harmful materials in spacecraft construction.

Limited consideration of social and cultural diversity: Although the report briefly mentions the importance of engaging the full diversity of America and its allies, a dissenting member may argue that it fails to fully explore the social and cultural dimensions of space development. They may advocate for a more comprehensive analysis of the ways in which space activities can be inclusive and representative of diverse communities, ensuring equal opportunities for participation and benefits.

Insufficient focus on long-term space strategy: While the report acknowledges the need for a long-term national space strategy, a dissenting member may contend that it lacks specific recommendations or actionable steps to achieve this goal. They may argue for a more detailed roadmap outlining how the United States can establish a clear vision and strategy for space development, including considerations for resource allocation, policy frameworks, and international cooperation.

In conclusion, a member of the organization responsible for this document might have principled, substantive reasons to dissent based on concerns related to international collaboration, the balance between national power and global cooperation, ethical implications, sustainability and environmental impacts, social and cultural diversity, and the lack of a comprehensive long-term space strategy.

RED TEAM CRITIQUE

Overall, the document is well-written and presents a comprehensive overview of the current state of the space industrial base in the United States. However, there are several areas where further analysis and clarification would strengthen the recommendations and insights provided.

Lack of Specificity: While the document emphasizes the importance of a clear national vision and strategy for space development and leadership, it fails to provide specific recommendations on what this vision and strategy should entail. It would be beneficial to include concrete steps and objectives that the United States should pursue to maintain its competitive edge in the space industry.

Inadequate Assessment of Competitors: The document briefly mentions increased competition in the space industry but does not delve into a detailed analysis of the competitive landscape. A more thorough examination of the strategies and capabilities of major spacefaring nations, such as China and Russia, would provide a more accurate understanding of the challenges that the United States faces in maintaining its position as a leader in this domain.

Limited Perspective on Diversity: While the document acknowledges the importance of engaging the full diversity of America and its allies in the space industry, it does not provide specific recommendations on how to achieve this goal. It would be valuable to explore strategies for increasing diversity and inclusion within the industry, such as promoting STEM education in underrepresented communities and implementing diversity hiring initiatives.

Insufficient Discussion of Regulatory Framework: The document briefly touches on the convergence of technology and the need for partnerships between the government and the private sector, but it fails to address the regulatory challenges associated with these developments. An in-depth analysis of the existing regulatory framework and recommendations for updating and streamlining regulations to foster innovation and growth would enhance the document's comprehensiveness.

Unrealistic Assumptions about Commercial Space Activities: The document suggests that commercial space activities have the potential to surpass government spending in the future, without providing a thorough analysis to support this claim. It would be more effective to present a balanced assessment of the opportunities and challenges associated with commercial space activities, including the risks of relying heavily on private sector investment.

In conclusion, while the document provides a solid overview of the current state of the space industrial base in the United States, it would benefit from more specific recommendations, a deeper analysis of competitors and regulatory challenges, and a more balanced perspective on commercial space activities. These additions would strengthen the document's recommendations and insights, making it more useful for policymakers and stakeholders in the space industry.

MAGA Perspective

This document is just another example of the establishment's attempt to push their globalist agenda in the name of space development. They talk about the importance of diversity and engaging with America's allies, but what they really mean is that they want to undermine American sovereignty and hand over our space capabilities to international organizations. We should be focused on putting America first, not worrying about what other countries think.

Furthermore, this document completely ignores the fact that the space industry is being flooded with competition from countries like China. Instead of addressing this threat head-on, they suggest vague recommendations and strategies that do nothing to protect American interests. It's clear that they don't have a real plan to maintain America's position as a leader in space.

The mention of partnerships between the government and the private sector is also concerning. While it may sound nice in theory, we've seen time and time again how these partnerships end up benefiting big corporations and leaving hardworking Americans behind. The government should be focusing on supporting American workers, not handing over control to profit-driven companies.

And let's not forget the mention of a long-term national space strategy. This is just code for more government bureaucracy and wasteful spending. We've already seen how ineffective government programs can be, so why would we want to double down on that?

Lastly, the document's emphasis on commercial space activities surpassing government spending is misguided. We shouldn't be relying on profit-driven companies to secure our position in space. We need a strong, centralized government approach to ensure that our space capabilities are protected and used to benefit all Americans, not just a select few.

PAGE-BY-PAGE SUMMARIES

AI-1   *Summary report on the state of the space industrial base in 2021, focusing on infrastructure, services, economic growth, and national security.*

AI-5   *This page provides a table of contents for a report on various aspects of space exploration and technology, including space mobility, policy and finance tools, space information services, and supply chain enablers. It also includes appendices with workshop participants, previous reports, and acronyms.*

AI-6   *The page discusses the importance of the United States' position in space and quotes President John F. Kennedy's commitment to leading in space exploration.*

AI-7   *The State of the Space Industrial Base 2021 report highlights opportunities and challenges for the United States in space, providing recommendations to advance economic, scientific, technological, and national security interests. It emphasizes the need for collaboration between government, industry, and academia to strengthen the space industrial base and promote innovation in space exploration.*

AI-8   *The State of the Space Industrial Base 2021 report outlines plans to accelerate the development of a vibrant space economy, establish standards and best practices, improve space architecture, enhance transportation and logistics infrastructure, and address energy demand and climate change.*

AI-9   *The report highlights the strength of the US space industrial base but warns of its fragility without strategic direction, workforce development, and adequate funding. It emphasizes the relevance of space to national priorities such as infrastructure, economy, climate change, talent diversity, and foreign policy. Major opportunities include setting a national vision, building a Cislunar economy, and leveraging commercial satellite capabilities for Joint All Domain Command and Control.*

AI-10   *The State of the Space Industrial Base 2021 report highlights the need for urgent action to address issues such as strategic guidance, resourcing, supply chains, and contracts in order to maintain American leadership in space. Recommendations include establishing a national space vision, incorporating the Moon into the Earth's economic sphere, increasing funding for space science and technology, and integrating commercial solutions for in-space logistics.*

AI-11   *The report highlights the importance of strategic direction, commercial contracts, domestic supply lines, and funding for research and development in order to maintain US leadership in space. It emphasizes the need to recognize the threat posed by China and take comprehensive action to advance in commercial, civil, and national security space.*

AI-12   *The State of the Space Industrial Base in 2021 highlights the need for cybersecurity and supply chain management in the space industry. Initiatives like the Hybrid Space Architecture aim to address these risks, but more needs to be done to foster collaboration, diversity, and engagement with universities. The page also mentions the importance of aligning space with national priorities, seizing opportunities for economic growth and climate change solutions, and addressing urgent issues such as strategic guidance and resourcing.*

AI-13   *Space is a valuable economic domain and strategic asset, and whoever establishes a sustainable presence on the Moon will have national preeminence and soft power advantages.*

AI-14   *The page discusses the importance of space in terms of national security, transportation, and international competition. It emphasizes the need for the United States to maintain its leadership in space and warns of the consequences if China were to surpass it.*

AI-15   *Space assets are considered infrastructure and play a vital role in various sectors. Treating them as such allows for new financial tools and attracts a wider range of investors, including* institutional investors seeking high returns. Q2 2021 saw a significant investment in space infrastructure, making it an attractive alternative investment option.

AI-16 Space infrastructure is critical to the functioning of our economy and society, supporting vital sectors such as finance, telecommunications, and power systems. The loss of GPS service alone would have a $1 billion per-day impact on the nation. Space should be recognized as critical infrastructure and efforts are being made to minimize risks to space systems. Additionally, NASA satellites have played a crucial role in our understanding of climate change.

AI-17 Space technology offers big ideas for climate solutions, including superior climate situational awareness, compliance monitoring, materials to support the green economy, avoiding industrial impact, controllable and reversible emergency climate interventions, and vast reservoirs of green energy. The US has the opportunity to lead in-space resource extraction, manufacturing, and transportation, while space-based information services can empower industries on Earth. Space may also become a destination for voluntary travel with a supporting hospitality industry.

AI-18 The page discusses the potential growth of the space industry, including job creation and economic benefits, and emphasizes the importance of US leadership in space to avoid ceding it to China.

AI-19 The page discusses the state of the space industrial base, highlighting the skilled labor and economic growth it brings. It emphasizes that space is America's competitive advantage, leading in exports and providing a positive trade balance. The global space industry is projected to reach trillions of dollars, with significant investment and potential for space tourism. Space technology offers innovation and a return on investment.

AI-20 The page discusses the economic impact of NASA activities and the growth of the global space economy. It also mentions changes in U.S. space policy under a new administration.

AI-21 China's space program has made significant advancements, including lunar sample return, Mars orbit, and the launch of a space station. The COVID-19 pandemic has affected the space industry, but investment is now increasing. Designating space systems as critical infrastructure is recommended for future protection.

AI-22 The commercial space sector rebounded after the COVID-19 pandemic, with major milestones achieved by Virgin Galactic, Blue Origin, and SpaceX. NASA awarded SpaceX the Artemis Human Lander System contract, and the Space Force's Rocket Systems Launch Program continued to prototype commercial launch services.

AI-23 The page discusses recent developments in the space industry, including successful satellite launches, investments in space infrastructure, and the significant venture capital funding attracted by space startups.

AI-24 SPACs have become a popular way to raise capital in the space industry, with $76.2 billion raised in 2020. Progress has been made on last year's recommendations, including the development of a vision for civil space and collaboration between the Department of Defense and NASA.

AI-25 The page discusses the importance of clarifying the role of the U.S. Space Force in protecting and enabling commerce in Cislunar space. It also mentions efforts by the U.S. government to stimulate the space industry, develop partnerships with allies, and establish international norms and collaborations.

AI-26 The page discusses the need for workforce development in the space industry and the importance of government support for commercial products and services. It highlights the challenges and risks associated with a lack of strategic guidance and international competition.

AI-27 Government space budgets are inadequate compared to the Cold War era, with military space science and technology (S&T) funding being particularly low. The U.S. must prioritize space

AI-28 Space is viewed as infrastructure for commerce and economic development, but policymakers are not adequately addressing the potential benefits of space industrialization and development. There is bipartisan support for US space leadership, with continuity across administrations in goals and programs.

AI-29 NASA's lunar exploration program, the Artemis program, has bipartisan support and aims to return humans to the Moon. The Space Force and Department of Defense are committed to monitoring and responding to any threats in Cislunar space. Think tanks are advocating for a broader strategic purpose and expanded responsibilities for the Space Force, including a focus on Cislunar space.

AI-30 The State of the Space Industrial Base 2021 highlights the success of new approaches in broadening the participation of non-traditional companies in national security innovation. However, challenges remain in diversifying procurement and connecting space policy with domestic and foreign priorities.

AI-31 The page discusses the need for space to be recognized as a 21st-century industry and a source of solutions for climate change. It highlights the importance of framing space as a national priority and the potential consequences of not doing so.

AI-32 The page discusses the growing importance of cislunar activities in the space industry, highlighting the need for new infrastructure. It also emphasizes the lack of awareness regarding the speed of innovation and competition in the industry.

AI-33 The page discusses the potential impact of fully reusable heavy lift rockets on global transport and space exploration. It also highlights the competition between countries in terms of infrastructure development and international standards in space.

AI-34 The page discusses the state of America's space industrial base and highlights the need for a clear vision and strategic direction. It addresses issues such as the lack of meaningful contracts, overclassification hindering innovation, and licensing bottlenecks faced by small companies.

AI-35 The page discusses the challenges and issues facing the space industry, including delays in satellite launches, the need for a central acquisition marketplace, capital misallocation, increasing competition from adversaries, and concerns about foreign ownership and influence.

AI-36 The State of the Space Industrial Base 2021 highlights the risks of China's influence in the space market, the lack of production targets for in-space industry, and the need for stronger international partnerships.

AI-37 Attendees recommend establishing a national vision and strategy for the economic development of space, incorporating the Moon into the Earth's economic sphere, and government involvement in infrastructure development to stimulate economic activity in the space industry.

AI-38 The page discusses the need for the United States to focus on developing infrastructure for space commerce, sustain funding for a hybrid space architecture, expand international norms of behavior in space, and increase funding for space science and technology.

AI-39 The page discusses the need to increase funding for space science and technology, reform policies to keep up with technological advancements, declare space a special economic zone, recognize space as critical infrastructure, include space in climate action plans, and prioritize space in supply chain planning.

AI-40 The page discusses the importance of integrating the Hybrid Space Architecture with Joint All Domain Command and Control (JADC2) for a more defendable space architecture. It also emphasizes the inclusion of commercial solutions in the DOD's in-space logistics infrastructure to enhance resilience and innovation.

AI-41 The page discusses the need to increase the percentage of commercial services acquisitions in the defense industry to promote innovation and benefit from faster industry cycles. It also highlights the importance of expanding the use and management of space commercial services within the Space Force to sustain capital investment and attract revenue.

AI-42 The page discusses the need for innovation and acquisition reform in the space industrial base, including increasing spending on commercial service contracts, creating a space commodities exchange, and shifting resources from SBIRs to OTAs. It emphasizes the importance of small businesses and the potential risks of not taking action.

AI-43 Investment in space beyond Low Earth Orbit (LEO) is needed for balanced growth, including in higher orbits and enabling technologies. The space industrial base should diversify and develop key components on-shore.

AI-44 The page provides an illustration of the Habitation and Logistics Outpost (HALO), a component of the Artemis Gateway set to launch in May 2024.

AI-45 The page discusses the need for a national vision and strategy for space industrial development in the United States, particularly in relation to creating a Cislunar economy and countering China's Belt and Road Initiative.

AI-46 Without a comprehensive national space vision and strategy, the US risks losing its space leadership and competitive advantage. A clear national space vision is critical for alignment, attracting partners, and driving innovation in the space industry.

AI-47 The page discusses the current state of the space industrial base in the United States, highlighting the need for national purpose, clarity, and strategic direction. It mentions the potential of space as a major shaper of national power and the challenges posed by rival nations. There is bipartisan support for U.S. space leadership, and a broader vision of the Space Force's strategic purpose is emerging.

AI-48 The page discusses the strategic importance of cislunar space and the need for the expansion of capabilities to protect American interests. Think tanks and experts have recommended a broader purpose for the Space Force, including offensive and defensive operations in space. The Space Force is committed to supporting space as a growing element of US national power.

AI-49 The page discusses the Congressional tasking to assess the strategic interest and capabilities of the United States and China in space exploration, as well as the potential environmental impacts of space-based resource extraction. It also highlights concerns about foreign investment, intellectual property theft, and China's efforts to control the US space industry supply chain. The page includes a table showing the lunar abundance of critical minerals essential to US economic and national security.

AI-50 Private investment in space led by entrepreneurs like Bezos and Musk is driving breakthroughs in cost, efficiency, and creative solutions. Their visions of expanding humanity into space and colonizing Mars have a significant impact on technology advances, public perception, and enthusiasm for the future of space. The US must match or exceed the ambitious goals set by China in space industrialization and economic development to maintain its space leadership.

AI-51 China's success in achieving its space goals has positioned it to attract new partners and investment, challenging US leadership. Space technology, such as GPS, plays a role in reducing global transportation emissions.

AI-52 The page discusses the need for a wealth-creation framework in space, sustaining a consistent vision across administrations, and the importance of democratic freedom and fair trade in space

AI-52 *activities. It highlights the competition with China and the potential consequences of ceding leadership in space.*

AI-53 *The page discusses the importance of the space industrial base and its potential contributions to addressing climate change, including improved climate situational awareness, compliance monitoring, materials for the green economy, reducing industrial impact on the biosphere, and controllable climate interventions.*

AI-54 *Space offers potential solutions for a green economy, including geo-engineering options and the exploration of green energy sources such as Helium-3 fusion. The United States should not overlook the opportunities presented by lunar mining and fusion technology.*

AI-55 *Space-Based Solar Power (SSP) is a promising solution for renewable energy and decarbonization. China leads in SSP research and aims to build a massive solar power satellite prototype and industrialize the Moon for solar power. Leadership in SSP would bring economic benefits, job creation, and control over strategic sectors in space. The US lacks a national program for SSP despite its potential.*

AI-56 *The page discusses proposals for a multi-billion Solar Power Satellite program as part of America's decarbonization strategy, with interest from various countries. However, the US lacks a coordinated effort and platform for such infrastructure, hindering progress in space solar power.*

AI-57 *The page discusses the need for the US to prioritize space development and settlement, including incorporating the Moon into the Earth's economic sphere and catalyzing space infrastructure and logistics. It also emphasizes the importance of the US government buying commercial products and services to drive demand.*

AI-58 *The State of the Space Industrial Base 2021 discusses the importance of the US having a national vision for space, the impact of partisan changes in administration, and the potential benefits of modular space design.*

AI-59 *The page outlines key actions and recommendations for the space industry, including updating goals, socializing the vision, establishing a task force, creating international institutions, constructing a defense alliance, and maintaining the cislunar economy.*

AI-60 *The page discusses the capabilities of the Starship in delivering spacecraft and materials to low Earth orbit, with logistics vehicles available for transportation to other destinations along the Space Superhighway.*

AI-61 *Space mobility and logistics are crucial for the success of the US space industry. The development of space infrastructure, including transportation and information systems, is essential for commercial viability and military advantages. Transitioning to modular approaches in spacecraft manufacturing will bring benefits in terms of flexibility and cost-effectiveness.*

AI-62 *Modular designs and standards in the space industry will lead to rapid technological advancements, a resilient space architecture, and reduced commercial risk. In-space logistics and plummeting launch costs are driving the growth of the industry. The terrestrial supply chain and resource availability are important considerations for space systems.*

AI-63 *The page discusses the need to secure materials and technologies for space systems, the importance of digital interoperability and cyber protections in space logistics, the success of interoperable satellite technologies, investments in in-space manufacturing, and the challenges faced by the U.S. space transportation industry.*

AI-64 *The page discusses the importance of interoperability standards in the space industry, the challenges of autonomous operations, the mismatch between R&D funding and space system costs, and the need for government investment to drive innovation and support space technology development.*

AI-65 The State of the Space Industrial Base 2021 highlights key points such as U.S. financial backing for infrastructure, growth in demand for broadband services, incentivizing emerging logistics providers, reducing barriers to entry for innovation, and the need for robust logistical support in space. The report also recommends placing propellant sources in space, increasing support to the space logistics industry, and utilizing commercial systems for logistics operations.

AI-66 The page discusses the concept of a space superhighway as a transformative and sustainable logistics infrastructure for space operations. It emphasizes the need for a whole-of-nation effort and a public-private partnership model to leverage commercial capabilities and support national security objectives. The U.S. Space Force can benefit from this infrastructure to enhance its missions.

AI-67 The page discusses the US government's engagement with academia and the adoption of modular capabilities in the space industry. It also highlights the importance of STEM investment, diversity, and sustainability in the space sector.

AI-68 Virgin Galactic's VSS Unity successfully reached space with Richard Branson and crew on board in July 2021.

AI-69 The page discusses the importance of space policy in sustaining and strengthening the US space industry. It highlights the need for alignment, prioritization, and visionary policy to unlock the economic potential of the space domain and maintain US preeminence in space.

AI-70 The US space industry has rebounded from the COVID-19 pandemic, attracting record levels of private investment. China is now a leader in certain space sectors. The US government has implemented strategies and regulations to support the commercial space industry, but further efforts are needed.

AI-71 Q1 2021 saw the formation of 8 new SPACs focused on the commercial space industry with over $21 billion in aggregate equity value. The National Space Council continues under Vice President Harris's leadership, providing a forum for future vision. The Intelligence Community establishes the IC Commercial Space Council and a Space Information Sharing and Analysis Center. The DOD pledges to abide by responsible behavior norms in space. Challenges and risks remain in moving forward efficiently in the space economy.

AI-72 The State of the Space Industrial Base 2021 highlights the need for government alignment, public interest, and a strong STEM workforce to sustain US space leadership. Advancements in space can support Biden's goals, such as reducing greenhouse gas emissions and creating new job opportunities. The space industry is projected to triple to $1.4 trillion within a decade, requiring investment in infrastructure and workforce development.

AI-73 Advancements in transportation modalities create economic opportunities, with early entrants like SpaceX and Rocket Lab USA dominating the commercial launch sector. The Netherlands' Golden Age demonstrates the potential for a power shift through increased access to space. China aims to surpass the US by seizing the first mover advantage in shaping the space economy.

AI-74 Early-stage space companies face investment challenges due to longer development timelines and the need for government contracts. The lack of clear demand signals from the government hinders industry growth. New entrants struggle to gain support and contracts, while the security clearance process favors established firms. Developing a STEM-powered workforce is crucial for future success in the space industry.

AI-75 The page discusses the state of the space industrial base in 2021, highlighting key points such as the U.S. vision for space, partnerships with emerging space powers, the potential for sustained human presence on the Moon, China's competition with the U.S., and the opportunity for the U.S. to fill a leadership vacuum in space.

AI-76 The page discusses key actions and recommendations for the space industrial base, including recognizing space as critical infrastructure, establishing a Space Commodities Exchange, adopting a readiness level framework, implementing a scheduled National Security Space Launch strategy, and creating a Chief Economist within the Department of Defense. It also highlights the economic benefits of GPS.

AI-77 The page discusses various strategies to address barriers and promote growth in the space industry, including modifying contract selection criteria, establishing a strategic propellant reserve, providing financial incentives for investors, and implementing a scholarship program for technical majors.

AI-78 The page discusses the state of the space industrial base in 2021, featuring an image of three Hawkeye 360 small satellites flying in formation.

AI-79 The page discusses the need for the Department of Defense to invest more in science and technology in the space information services industry. It also highlights the importance of embracing competition and commercial services, as well as the potential impact of remote sensing, AI, and connectivity on intelligence gathering.

AI-80 The page discusses the state of the space industrial base in 2021, highlighting the need for a Hybrid Space Architecture and the expansion of commercial communications networks. It also mentions the progress of the space industrial base and the challenges faced by the U.S. government in embracing commercial space infrastructure.

AI-81 The page discusses the state of the space industrial base in 2021, highlighting the launch of IoT satellites, advancements in space-based position, navigation, and timing, challenges in government acquisition of commercial weather monitoring, and the growth of commercial assets in space situational awareness.

AI-82 The State of the Space Industrial Base in 2021 highlights challenges in acquiring and utilizing commercial space data for defense and intelligence purposes. Issues include outdated acquisition regulations, difficulties with security clearances, limited purchasing options, and constraints on exporting data. These obstacles hinder the effectiveness and competitiveness of US companies in the global market.

AI-83 An undergraduate student discovered 120 ICBM silos under construction in China using commercial remote sensing services, confirming the findings with high-resolution imagery. Commercial remote sensing and analytics are valuable for detecting anomalous behavior and improving situational awareness.

AI-84 The State of the Space Industrial Base in 2021 highlights key developments such as the establishment of a commercial space services acquisition office, adoption of hybrid space architecture standards, and improvements in cloud analytics. Recommendations include funding for a Hybrid Space Architecture program and expanding the mission of the USSF Space Systems Command.

AI-85 The page discusses the need for increased funding and support for commercial space capabilities, including the development of integrated systems and tools, standardized communication technologies, and a space internet consortium.

AI-86 Space is now accessible to people of all backgrounds and ages, thanks to American innovation and entrepreneurship. This is demonstrated by figures such as Richard Branson, Jeff Bezos, and the crew of Inspiration4.

AI-87 The United States is on the verge of a significant shift in space exploration, with increasing involvement from the private sector. The goal is to enable and secure commercial human ventures in space, as well as maintain strategic relationships to ensure the US remains a leading player in space exploration.

AI-88 The US space industry is entering a new era with the retirement of the Space Shuttle and the emergence of commercial space travel. SpaceX and Boeing are transporting NASA astronauts to the International Space Station, while other companies like Blue Origin and Virgin Galactic are offering suborbital flights for personal space travel. NASA is also pursuing the Artemis program to return humans to the Moon by 2024, with SpaceX leading the way in lunar landings and logistical supply. The development of robotic systems and nuclear thermal propulsion are key

AI-89 Government-industry partnerships are driving the commercialization of space, with plans for commercial astronaut visits to the ISS and private trips to space. Companies are also developing sensor networks to improve space domain awareness. The key challenge is to make space travel safer, desirable, and affordable to support sustained human presence in space.

AI-90 The page discusses the challenges and advancements in the space industry, including communication capacity, autonomy, energy supplies, and key milestones such as personal human transport to near space, lunar missions, and decreasing launch costs.

AI-91 The page discusses the need for human presence in space for manufacturing and resource extraction, as well as the importance of propellant production on the Lunar surface. It provides key actions and recommendations for short-term, mid-term, and long-term payoffs in the space industrial base.

AI-92 The page discusses the Space Infrastructure Dexterous Robot (SPIDER), a technology demonstration for NASA's OSAM-1 mission.

AI-93 The page discusses the growing importance of space outside of Earth's orbit and the opportunities it presents for commercial development and military operations. It highlights the need for superior capabilities in communication, movement, and sustainment in order to dominate the future development of this domain.

AI-94 Cislunar orbits are complex and unpredictable, making it difficult to describe and visualize trajectories. The global supply chain for space components is struggling to keep up with demand, leading to delays in Cislunar projects. China is expanding its presence in Cislunar space, while private sector investment remains focused on platform technologies and activities in low Earth orbit.

AI-95 The page discusses the importance of communication and navigation infrastructure in Cislunar space, the limitations of current propulsion systems, the growth of new markets in on-orbit servicing and manufacturing, and the challenges faced by the space industrial base in responding to increased demand.

AI-96 Satellite integration and testing facilities, regulatory/licensing approvals, and a lack of defined market and interface standards are limiting factors in space technology development. Small businesses face barriers in government-sponsored technology development, and there is an aversion to technical risk. Slow growth may threaten US dominance in the space industry.

AI-97 The page discusses the potential futures of the Cislunar space industry, including scenarios such as NASA domination, competition for lunar resources, and the collapse of support for Cislunar activities. It also highlights key indicators of progress in areas such as communications, mobility, servicing and assembly, manufacturing, and lunar outpost construction.

AI-98 The page discusses key actions and recommendations for the space industrial base. It suggests continuing procurement of commercial products and services in Earth orbit and Cislunar space, committing to Cislunar infrastructure, government buying Lunar internet/data service, supporting advanced propulsion RDT&E, using prizes for DOD-relevant technologies, removing barriers for small businesses, and re-examining risk stance to encourage innovation.

AI-99 Space industrialization is a reality with companies advancing technology for global markets. The space domain offers opportunities for first-mover advantage and a robust US industrial base is crucial for national security and military advantage.

AI-100 The page discusses the importance of the space industry and the need for government investment to create a self-sustaining and scalable industry. It suggests tying in climate change, power, resources, and on-orbit assembly and manufacturing into broader policy initiatives. Future work is recommended to develop specific proposals for immediate action.

AI-102 A list of individuals and organizations involved in the space industry, including companies, government agencies, and research institutions.

AI-104 Summary: State of the Space Industrial Base 2021 conference attendees include representatives from various organizations in the space industry, discussing topics related to missile defense, innovation, product transitions, NASA, Air Force, space technology solutions, and more.

AI-105 The page discusses previous reports and key recommendations regarding the state of the space industrial base. It emphasizes the need for a long-term national space strategy, the development of essential capabilities and technologies, and continued investment in science and technology.

AI-106 The page discusses the future of space in 2060, highlighting the complexity and diversity of state and non-state actors. It emphasizes the challenges and opportunities of commercial space and the need for alliances and partnerships. Another report focuses on sustaining US leadership in space, providing recommendations for policymakers and industry leaders to integrate national power and stimulate the space industry.

AI-107 The page discusses the need for a grand strategy for space, the importance of a strong US Space Industrial Base, and potential future missions and advancements in space science and technology.

AI-109 The 2021 DIU Space Portfolio Company Survey reveals that many companies in the Space Industrial Base face challenges due to unpredictable government demand, export controls, and supply chain issues. Most respondents are small, non-traditional defense contractors heavily reliant on U.S. revenue.

AI-110 The State of the Space Industrial Base 2021 report found that the majority of surveyed U.S. commercial space companies are small businesses with potential for growth in the new space market.

AI-111 The State of the Space Industrial Base 2021 report shows that hiring difficulty during the COVID-19 pandemic has not changed significantly overall, but there are variations by state. Companies in New Mexico and Florida have struggled to hire new employees, while Colorado and New York have found it easier. Financially, most companies in the space industry are better positioned in 2021 compared to 2020, with increased investment and government contracts providing stability.

AI-112 Approximately 56% of surveyed companies have received SBIR grants, which are used for R&D and capital expenditures. However, the application process is difficult and confusing, leading to a "valley of death" beyond Phase 2 where companies struggle to connect their technology to DOD programs.

AI-113 A quarter of non-traditional vendors in the space industry are considering SPACs as a viable exit strategy, primarily for immediate access to capital and freedom from recurring fundraising activities. Some commercial space companies with DOD contracts have achieved unicorn status.

AI-114 Two thirds of respondents in the space industry are experiencing supply chain delays, with 39% facing delays of eight weeks or more. The COVID-19 pandemic has exacerbated the existing tightness in global semiconductor supply, leading to shortages of critical components such as microchips. There is a renewed demand for radiation-hardened electronics in the space industry.

AI-115 Government demand signal and ITAR regulations are seen as hurdles to future growth in the space industrial base, with challenges including access to government contacts, security clearances, and lengthy contracting processes. ITAR is particularly problematic due to its impact on market access and competition from "ITAR-free" competitors.

AI-117 The Quality Analytics report highlights the high launch costs as a barrier to the space industry's expansion. However, over the past decade, launch costs have significantly decreased, paving the way for a new space economy. Commercial activities are expected to surpass government spending, with a focus on delivering services and the emergence of new industries in low Earth orbit and Cislunar space. The report emphasizes the need for investment in space infrastructure and the adoption of new engagement models to stay ahead of adversaries.

AI-118 The space industry has unique challenges including long development cycles, limited testing capabilities, and high costs. The government should focus on buying existing products, engaging with small-to-mid-sized companies, and considering international partnerships to drive innovation and investment in the space industry.

AI-119 The page discusses the importance of a Hybrid Space Architecture, which integrates new smallsat capabilities with traditional government space systems. This architecture aims to improve deterrence, resilience, and information advantage in space while promoting risk distribution, faster operations and innovation, improved interoperability, and leadership in the space economy. It leverages various technologies and approaches such as secure communications, artificial intelligence, distributed ledgers, cloud infrastructure, commercial space manufacturing efficiencies, and rapid acquisition mechanisms.

AI-121 This page provides a list of acronyms and abbreviations related to the state of the space industrial base in 2021.

AI-123 The page provides a list of abbreviations and acronyms related to the space industry and government organizations.

NOTABLE PASSAGES

AI-6 "The United States of America has no intention of finishing second in space. This effort is expensive - but it pays its way for freedom and for America." - PRESIDENT JOHN F. KENNEDY, 1963

AI-7 "While these recommendations do not represent the official position of the United States Space Force, NASA or any other branch of government, they are extremely valuable inputs for consideration."

AI-8 "Our vision is a bright and bountiful space future that is safe, secure, sustainable, synergistic, and successful."

AI-9 "While the pace of innovation and investment in the U.S. is at an all-time high, participants cautioned that this will not be sustained without strategic direction, robust adoption of commercial space capabilities expressed in meaningful contract opportunities, strategic workforce development, attention to fragile domestic supply lines, and addressing the anemic funding to prototype, validate and accelerate the adoption of innovative and disruptive space capabilities for national security."

AI-10 "A robust/competitive U.S. Space Industrial Base is essential to maintaining the U.S. as a preeminent space power, but its competitive advantage is threatened by increasing globalization of space industrial capabilities. Other nations are aggressively expanding their space industrial bases. To meet these challenges, the USSF must lead and develop an all-of-government strategy to partner with the U.S. Space Industrial Base, taking advantage of present, commercial capabilities while stimulating future commercial capability development for U.S. space military needs."

AI-11 "Leadership in space exploration has a real political meaning...Failure in that leadership means inevitably falling into the status of a second-class nation with the heavy costs to our way of free enterprise which subjugation to others would involve." - LLOYD V. BERKNER, 1960

AI-12 "Participants asserted that major opportunities exist to set a national vision, to build an inclusive Cislunar economy, to set key technical and behavioral standards which advantage democracies, to leverage existing U.S. and allied commercial space capabilities for a hybrid network architecture enabling the Joint All Domain Command and Control (JADC2), a space internet, a space superhighway for logistics, and to apply space solutions to Presidential and Secretary of Defense priorities for climate change."

AI-13 "Whoever is first to establish a sustainable presence on the Lunar surface will retain both the perception and reality of national preeminence." - DR. BHAVYA LAL, NASA, 2021

AI-14 "If led by the U.S., it offers opportunities to deepen its alliance with international partners, including possibly Russia, while expanding future collaborations to include new partners in the Middle East, Asia, Africa, and South America. If this competition is won by China, they will leverage the substantial soft power it would gain from Lunar preeminence to achieve a wide variety of national security, economic, and diplomatic/political objectives. The global perception of the United States will be substantially wounded, feeding a narrative of American decline and

AI-14 (cont.) *China's ascendance. China conducting these activities in partnership with Russia would have even graver implications for national security."*

AI-15 *"Space is Infrastructure - The assets we have in space are best thought of as infrastructure. Space assets underpin nearly every sector of our society. They synchronize our power grid. They synchronize, coordinate and secure our financial transactions. They connect our cities and rural areas, providing long-distance communications for television, radio, telephony and broadband internet. They supply weather, traffic and logistics data to enable city planning, agriculture, public health and transportation. They provide advance warning of a broad range of hazards including extreme weather events to safeguard lives, crops, and critical infrastructure."*

AI-16 *"The loss of GPS service would average a $1 billion per-day impact to the nation."*

AI-17 *"""Space technology offers big ideas for climate solutions in six distinct categories:*

- *Superior Climate Situational Awareness*

- *Compliance Monitoring*

- *Materials to Support the Green Economy*

- *Avoiding Industrial Impact*

- *Controllable and Reversible Emergency Climate Interventions*

- *Vast Reservoirs of Green Energy*

AI-18 *"Financing [Space Solar Power Satellites] a little-known element of NASA's union-built, clean energy technology through that infrastructure bill is the fastest, most painless way to accomplish, and pay for, the Green New Deal (GND) and President Biden's goal to Build Back Better... It will also bring unexpected "space benefits" to Native American, Black and Latino populations across the nation." - INT'L ASSOCIATION OF MACHINISTS AND AEROSPACE WORKERS, 2021*

AI-19 *"The global space industry is estimated by the Space Foundation to have been $446.9 billion in 2020. This $447 billion economy is 55% larger than a decade ago, and part of a five-year trend of uninterrupted growth. Projections for 2040 range from 1-1.5 trillion (Goldman Sachs $1.1 trillion, Morgan Stanley $1.1 trillion, U.S. Chamber of Commerce, $1.5 trillion), and projections for 2050 range from $2.7 trillion (Bank of America Merrill Lynch) to $10 trillion (China Academy of Space Technology). A total of $177.7 billion of new capital investment has been added to the commercial space industry in the last*

AI-20 *"NASA activities in the U.S generated $64.3 billion in economic output -- a return on investment of around 200%. NASA activities also supported 312,630 jobs in the U.S. in FY 2019, nearly 50,000 of which are in scientific research and development disciplines. That R&D pays off, with $14.2 billion of NASA's generated economic output in the scientific research and development sector and documented cases of over 2,000 spinoff technologies developed by NASA or developed with NASA assistance. Simply put, space dollars are dollars that result in a larger U.S. economy."*

AI-21 *"America's competitors continued their effortss to assume leadership. Just in the interval since the last report - one year -the Communist Party of China (CPC) succeeded in its announced goals to execute a Lunar sample return, achieve Mars orbit, land, and deploy a rover, launch and crew its competitor space station and test suborbital transportation systems. China also experienced a surge in commercial investment following space being designated 'new infrastructure' along with 5G. China opened its space station for international experiments via the United Nations. China and Russia announced a joint Lunar base and released a user's guide for nations interested in partnering.*

AI-22 *"The commercial space sector made an amazing rebound following the unforeseen tightening of the capital markets during the COVID-19 pandemic. The U.S. must be more resilient in the future."*

AI-23 "Over the last 10 years, there has been $199.8 billion of equity investment in the U.S. and China, which collectively accounts for 75% of the global total investment in space. After four quarters of declining deal volume, investors deployed another $9.8 billion in space companies in Q2 2021, which is the fourth largest quarter on record for total space investment. Despite only two space company SPACs closing in Q2 2021, this was the largest quarter on record for space infrastructure investment. With another $4.5 billion invested in the quarter, total infrastructure investment is now on pace to beat the previous annual record year of $9.1 billion set last year."

AI-24 "SPACs allow qualified investors to create a 'blank check' company which goes public with an intention to acquire and consolidate a start-up with high future earnings potential, and merge it into the publicly traded company. This event results in an infusion of capital into the company while providing an exit opportunity to early investors."

AI-25 n/a

AI-26 "Despite record levels of private investment in a record number of U.S. space companies offering a steady parade of new products and services, success is hinged on shaky support by the U.S. Government as a consumer of commercial products and services. The notable exception here is NASA, which continues to strengthen both its direct investment in, and procurement of, commercial space offerings. According to participants, a lack of strategic guidance and clarity; the inability to reward private capital investment with meaningful contracts -- to 'buy commercial;' a fragile supply chain; and strong international competition and predatory practices by adversaries means that failure to take action puts the gains so far at risk."

AI-27 "Recent analysis suggests that the PRC is already at parity or outspending the U.S. defense budget. Overall DOD S&T is reported to be just 70% of Cold War levels at just 1.9% of the DOD budget."

AI-28 "Future investments in transportation and industrial facilities to support an in-space Cislunar econosphere are likened to our nation's past investments in its canal system, Transcontinental Railway, national highway system, aviation infrastructure, and internet."

AI-29 "NASA's 21st century lunar exploration program will make new discoveries, advance technologies, and show us how to live and to work in another world. For that to be possible, the space program needs constancy. That's why NASA must be a nonpartisan agency, and why the Artemis program has bipartisan support. For the first time in more than 50 years, NASA will return humans to the Moon. We will go in a way that reflects the world today, with government, with industry, and with international partners in a global effort."

AI-30 "If the U.S. is to retain its technological leadership and compete globally, it must diversify its portfolio and increase the total percentage of commercial procurements as prescribed by Congress. This is keeping with the adage that the U.S. Government should 'buy what it can, and only build what it must.' In doing so, more buying power is achieved by negating design, sustainment, and improvement costs."

AI-31 "What's at Stake - is no less than whether the largest geographic zone of human activity is one of democratic freedom and fair trade, or an autocratic exclusion zone. Will citizens of the world conduct their activity under a presumption of freedom, human rights, rule of law and a rules-based-order, or will they be mere extensions of a totalitarian state using rule-by-law? Will the material resources of outer space-a million-fold what is on Earth accrue to totalitarian powers and their resource-nationalist tendencies, or will they accrue to U.S. and like-minded nations where a balance of power that favors freedom prevails? Will the industries and jobs and partnerships of the 21st century, including leadership on climate change and green technologies accrue to the

AI-32 "Few are aware that in the last decade launch costs have dropped ten-fold, the number of active satellites in orbit have more than tripled."

AI-33 "They are not accounting for a 'Starship Singularity' where fully reusable heavy lift rockets will enable global transport in under 30 minutes, daily launches to orbit of 100 metric tons for $2-5 million at <$100/kg. These same vehicles are being designed to refuel on orbit and to be able to land on all rocky planets and moons in the solar system."

AI-34 "America's space industrial base is the equivalent of a superpower. Its talent and energies are vast. But it is like a disorganized team waiting on a team captain to call the play, coordinate and motivate the team to compete."

AI-35 Transitioning the 'Valley of Death' - Although signiflightcant progress has been made in flightnding and maturing commercial solutions through AFRL, DIU and others, the limiting factor is the low level of subscription by acquisition services or programs of record which are often tied to 'requirements'-based acquisition. The DOD needs a central acquisition marketplace for commercial products and services drawn from the commercial sector.

AI-36 "The lack of clear production targets for in-space industry for commercial partners associated with the Artemis Program limits its impact as a tool for space resource and economic development. A strategic call for the creation of in-space logistics infrastructure, both physical and digital, would otherwise benefit civil, commercial and military/national security space."

AI-37 "The United States is the firstrst country to have private companies taking private passengers to space. This is a moment of American exceptionalism. That's how we see it...It will be the ingenuity of all of our commercial partners to help us continue advancing to the next stage of our nation's space exploration." - JEN PSAKI, White House Press Secretary, 2021"

AI-38 "The U.S. must accelerate and sustain this activity to thwart China's ambitious goal to dominate the Space Internet."

AI-39 "The fact that space systems constitute infrastructure critical to our way of life and prosperity is now broadly accepted, yet it is not formally recognized. Recognizing space as critical infrastructure opens new tools to sustain and grow the sector."

AI-40 "Enable the Space Superhighway by Including Commercial Solutions for In-space Logistics Infrastructure - As USSF articulates its architecture for in-space mobility and logistics, it is critical to include commercial solutions from the start. This effort should be done in close collaboration with NASA. In terrestrial modes of logistics, the ability to make use of civilian ports, civilian fuel, and civilian interfaces is a significant force multiplier -- the same is true in space. A logistics architecture that leverages commercial solutions will be more resilient, scalable, provide greater volume and reach, and see faster innovation because of private capital investment and its broader and more scalable economic impact."

AI-41 "The current policies, budgeting structure, and lack of procurement innovation incentives (plus perceived risk) contribute to this low level. Participants judged that a commitment to more agile and rapid innovation requires policy and incentives that drive toward a goal of 20% non-traditional commercial service acquisitions. This would greatly ameliorate the various technology valleys of death and enable the DOD to benefit from the much faster innovation cycles in industry."

AI-42 "The side that wins in the future is the side that has the greatest situational awareness and acts most quickly."

AI-43 "Balanced Growth Requires Investment Beyond LEO - The innovations in venture capital and SPACs have put the Low Earth Orbit (LEO) economy on solid ground with a tremendous diversity of space access options and space information services. However, we do not yet have a MEO, HEO, GEO, Cislunar or Lunar economy that is new space-oriented. This is despite the fact that there is clear interest from the government in expanding activity into these areas, a willingness to extend security into these areas, and no lack of capable start-ups with mature-enough technology *seeking funding. To maintain growth, we need to diversify investment beyond LEO, expanding investments both in scale and to higher orbits. We want scale with the U.S.*

AI-44 n/a

AI-45 "The U.S. must develop and execute a grand strategy for space recognizing space's importance and enhancing our advantages. This strategy must encompass the near-term future, with space oriented as a source for augmenting terrestrial power, and the long-term future, encompassing space across the Cislunar expanse and beyond as a domain in itself for human action." - USSF SPACE FUTURES WORKSHOP REPORT, 2021

AI-46 "Participants judged that without an integrated, comprehensive national space vision and strategy, U.S. space leadership and competitive advantage are at risk. Without a coordinated, consolidated national vision and policy there is insufficient demand signal to U.S. research and development agencies and the private sector. There is no national vision with articulated goals to 2050. There is also no national level vision that provides clear, time-specified goals which tie space power milestones to a broader long-term competitive industrial strategy and vision."

AI-47 "The U.S. cannot ignore the potential of space as a major shaper of our present and future national power and the power of our rivals and adversaries. History shows nations who ignore new or expanding domains of human endeavor suffer for it. Other competitor and allied nations recognizing this potential are moving aggressively to position themselves in this future space world. The U.S. has advantages (historical, economic, political, and intellectual) we must exploit to meet these challenges."

AI-48 "we are opening our aperture to keep pace with our nation's expansion into the Cislunar region, to the Moon, Mars and beyond."

AI-49 "The NDAA tasking is clearly not only about exploration, but asks specifically for 'a comprehensive assessment between the United States and China' not only of human exploration and spaceflight, current and future space launch capabilities, but also to assess 'the strategic interest in and capabilities for Cislunar space' and 'the viability and potential environmental impacts of extraction of space-based precious minerals, on-site exploitation of space-based natural resources, and the use of space-based solar power.' The NDAA task echoes many of the concerns in this report regarding the extent of foreign investment in the commercial space sector; theft of United States intellectual property; efforts by China to seize control of critical elements of the United States space industrial supply chain and United States space industry companies, and

AI-50 "Bezos dreams of 'moving all heavy industry off Earth' so that Earth can be 'zoned residential and light industry' and 'a world for his great-grandchildren's grandchildren where humanity moves out into the solar system.' Musk aspires to colonize Mars and make humanity a multi-planetary species. They both speak of millions of people living in space, and of scales far more ambitious than articulated in national vision or policy. Some may question their idealism in favor of other rationales, but their impact on U.S.' and allies' space technology advances and U.S. commercial space advantage continues to be dramatic - even more so their impact on general public perception and enthusiasm for the future of space and for continued U.S. space leadership

AI-51 China has expended significant political capital committing to big audacious goals to aid its public diplomacy, soft power, attractiveness as a partner, and international prestige. The PRC mobilized its national energies toward their fulfillment of their announced goals on time, and this year it succeeded in efforts announced over a decade ago: Lunar sample return, achieve Mars orbit, land, and deploy a rover, launch and crew it's competitor space station and test suborbital transportation systems.

AI-52 "What's at Stake - is no less than whether the largest geographic zone of human activity is one of democratic freedom and fair trade, or an autocratic exclusion zone. Will citizens of the world conduct their activity under a presumption of freedom, human rights, rule of law and a rules-based-order, or as mere extensions of a totalitarian state using 'rule-by-law'? Will the material resources of outer space-a million-fold what is on Earth accrue to totalitarian powers and their resource-nationalist tendencies, or will they accrue to U.S. and like-minded nations where a balance of power that favors freedom prevails? Will the industries and jobs and partnerships of the 21st century, including leadership on climate change and green technologies accrue to the

AI-53 "Strategic advantage in space compounds over time analogous to compound interest in a bank. Therefore, initial conditions matter, and create a path dependence for all future participants. The nation that emerges as the leader will set the precedents that condition the system, determining the rules of the playing field. Action therefore is urgent if we desire to retain U.S. leadership and secure for our children and grandchildren a second American century."

AI-54 "The Moon could serve as a new and tremendous supplier of energy and resources for human beings. This is crucial to sustainable development of human beings on Earth...Whoever first conquers the Moon will benefit first."

AI-55 "It means constructing large orbital solar farms that collect the intense solar energy above the clouds and where there is no night, and transmit the energy wirelessly to the ground. If constructed using the materials of the Moon or asteroids, these power stations could scale to all global demand many times over, with the International Academy of Astronautics assessing that 'annual employment on the order of 5,000,000 individuals might be realized eventually.' This is an idea invented in America by Dr. Peter Glaser, and recommended for action by the Pentagon study group over a decade ago."

AI-56 "United States the only major space faring nation whose national space agency does not have a serious plan to develop a SSP platform...Given SSP's benefirstts and the interest in the technology from most other space agencies, it's puzzling that policymakers in the United States have not prioritized SSP R&D." - PROGRESSIVE POLICY INSTITUTE, 2021

AI-57 "Make Space Development and Settlement the U.S. National Vision - The full mobilization of America's diverse talent set requires big audacious goals which cause the nation to stretch, aid public diplomacy, and create the perceptions of the U.S. as a vibrant attractive partner. Space offers solutions for tackling climate change -- not only in monitoring and modeling but in scalable energy solutions to create a green space power grid and source the materials for an electric economy."

AI-58 "The U.S. must develop and execute a grand strategy for space recognizing space's importance and enhancing our advantages. This strategy must encompass near-term terrestrial-focused power and a long-term focus on Cislunar expansion and beyond as a domain in itself for human action." - USSF SPACE FUTURES WORKSHOP REPORT, 2021

AI-59 "The USSF is committed to its broader strategic purpose to support space as a growing element of U.S. national power ...the Space Force will 'be there' wherever U.S. commercial and strategic interests and activities expand."

AI-60 n/a

AI-61 "Logistics is the bridge between the economy of the Nation and the tactical operations of its combat forces. Obviously then, the logistics system must be in harmony, both with the economic system of the Nation and with the tactical concepts and environment of the combat forces." - REAR ADMIRAL HENRY ECCLES, 1959

AI-62 "In space, modular designs will allow rapid introduction of new technologies, correction of failed components, and less frequent disposal. Launching modules rather than complete spacecraft will shift the launch process from chartered to scheduled. More companies will be qualified to produce space hardware, expanding and enhancing the supply chains and industrial base. Developing the appropriate incentives to ignite the modular transformation should start now. Combined with a mesh network for in-space communication, modularity will lead to a highly resilient DOD space architecture."

AI-63 "In the past year, multiple instances of tipping and queuing satellites to utilize the best aspects of our commercial satellite technologies demonstrated the power of interoperable logistics within LEO. Broad coverage from large EO/IR constellations discover areas of interest using wide area analytical technologies. These broad area coverage satellites are able to provide exact coordinates to satellites with Synthetic Aperture Radar (SAR) capable of extremely high-fidelity imagery. This digital logistics chain has proven itself in rapidly tasking and imaging hot areas of interest, to include illegal fishing, smuggling, sanctions violations, military movements and construction, and work camp construction."

AI-64 "Interoperability standards and frameworks must be established early and with broad industry engagement so that the government doesn't adopt a standard that the commercial industry does not adopt, creating a rift in technological development. To alleviate these challenges, the government must carefully and collaboratively forge enabling interoperability standards early and upfront, then indicate what technologies are vital for national application via investment and the purchase of services, but allow commercial consensus to drive and provide feedback on how to best deliver these requirements."

AI-65 "Between 2015 and 2025, we have an opportunity to put a fleeteet on another sea. And that sea is space." - GENERAL CHARLES KRULAK, USMC, 1997

AI-66 "The Space Superhighway concept has the potential to transform American space operations from a disposable, vulnerable, aging fleet to a vibrant, dynamic, and sustainable system. It promises economic impact on the order of our Interstate Highway System, the Transcontinental Railroad or other large infrastructure investments."

AI-67 USG enables a sustainable climate future by adopting modular logistics approaches that provide more frequent and affordable monitoring of rising sea levels, illegal fishing, and oil spills. As our technologies and sensors increase in capability and our data services track environmental changes, the USG can be on the forefront to preserving the planet for future generations.

AI-68 n/a

AI-69 "To all you kids down there, I was once a child with a dream looking up to the stars - now I'm an adult in a spaceship with lots of other wonderful adults looking down to our beautiful Earth. To the next generation of dreamers - if we can do this, just imagine what you can do." - SIR RICHARD BRANSON, CEO Virgin Galactic, 11 July 2021

AI-70 "The U.S. space industry was initially shaking at the start of the COVID-19 pandemic but quickly bounced back, attracting record levels of private investment in 2021."

AI-71 "One can imagine a self-reinforcing virtuous cycle of development that would support the space economy. But one can also reasonably doubt that such an ideal path will be realized easily or without some nudges along the way. Limits on or asymmetries of information, the high level of risk inherent in space and the challenges of capturing surplus from such complementarities will make it difficult to move forward on the most efficient path-or even to move forward at all." - MATTHEW WEINZIERL, Harvard Business School Professor, 2018.

AI-72 "The space industry is projected to triple to $1.4 trillion within a decade. A healthy space infrastructure supported by a STEM workforce must be put in place to capture a dominant share of this future economic growth."

AI-73 "Throughout human history, the greatest advancements in economic opportunity can be intrinsically linked to the introduction of new transportation modalities that have forever changed the economic and military influence in state affairs. History also reveals that such opportunities are fleeting. In other words, there is a significant first mover advantage for those

AI-73 (cont.) "who recognize and establish early entrance and leadership in new and emerging transportation markets."

AI-74 "If we do not make the strong eeffortort now, the time will soon be reached when the margin of control over space and over men's minds through space accomplishments will have swung so far on the Russian side that we will not be able to catch up, let alone assume leadership." - LYNDON B. JOHNSON, Vice President, 1961

AI-75 "China becomes the leader by out-competing the U.S. in providing lower-cost, higher-capability space services to U.S. allies and partners."

AI-76 "The exploration of space will go ahead, whether we join in it or not and it is one of the great adventures of all time and no nation which expects to be the leader of other nations can expect to stay behind in this race for space." - JOHN F. KENNEDY, President, 1962

AI-79 "The secret of war lies in the communications." - NAPOLEON BONAPARTE

AI-80 "The U.S. should compete by making unclassiflighted commercial space services available to allies and partners."

AI-81 "Government acquisition of commercial weather monitoring and prediction has been stifled by international legislation, especially World Meteorological Organization (WMO) Resolution 40, requiring governments to freely share all meteorological data (thus undermining the business plan of any commercial provider). The new U.S. Government administration indicated that climate change will be a high priority over the coming years, signaling a potentially fertile market for commercial sensing despite its own related WMO Resolution on sharing."

AI-82 "Warflightghters, including eflightectively all COCOMs, have made strident demands for more releasable data. Furthermore, General Raymond has repeatedly called for a hybrid space architecture to provide resilience to U.S. space assets. However, these goals are facing signiflightcant systemic and structural roadblocks."

AI-83 "Student discovers 120 ICBM silos under construction in China using commercial remote sensing services"

AI-84 "The GEOINT singularity will occur as sensors, on-orbit computation, AI/ML analytics and ubiquitous connectivity converge."

AI-85 "As the Administration undertakes its review and update of national space policies, as well as defense and intelligence strategies, there is an opportunity to realize longstanding policy to 'rely to the maximum practical extent' on commercial capabilities. This would start with taking a fresh look at the role that commercial space plays in hybrid architectures. Commercial space can also contribute to experimentation with new operating concepts, new distribution networks, and new pathways to share information with allies and partners, as well as being a prime candidate for exploring new acquisition models that are better suited to the pace of commercial technology innovation."

AI-86 "Space is finally open to all: young and old, rich and poor and from all walks of life thanks to American ingenuity and entrepreneurship."

AI-87 ""The goal isn't just scientifirstc exploration... it's also about extenting the range of human habitat out from Earth into the solar system as we go forward in time... In the long run a single-planet species will not survive... There will be another mass-extinction event. If we humans want to survive for hundreds of thousands or millions of years, we must ultimately populate other planets.... I'm talking about that one day, I don't know when that day is, but there will be more human beings who live off the Earth than on it." -- HON. MICHAEL D. GRIFFIN, 2005"

AI-88 "The U.S. is On the Threshold of a New Era in Human Presence in Space - Almost ten years after the retirement of the Space Shuttle, Space Exploration Technologies (SpaceX) launched on 30

May their DM-2 mission carrying two NASA astronauts to the International Space Station (ISS) for a 3-month mission. This flightight is the culmination of the NASA Commercial Crew program which helped create SpaceX's Dragon 2 and Boeing's Starliner for the purpose of transporting NASA astronauts to ISS and back."

AI-89 "Make it Safer - Drive towards space travel being as safe as air travel is today. Make it Desirable - Drive towards making travel to/from and habitation in space easy and enjoyable. Make it Affordable - Drive down costs towards a price point <$100/kg that will enable millions of people to be able to afford to travel to space."

AI-90 "Expanding Autonomy - Logistics missions that support human activity cannot affordably or safely rely on continuous control and monitoring from Earth. Autonomous systems that can be validated for safety will be required to expand to the coming level of in-space operations, both human and robotic."

AI-94 "Cislunar Orbits are Complicated - The two-body (Keplerian) assumption that works quite well in describing orbits of satellites below GEO does not translate well to the Earth-Moon system. Here, three-body effects can cause large deviations from traditional orbit elements (e.g. eccentricity, inclination, etc.) and are not well described by the Two-Line Element (TLE). The result is that trajectories in Cislunar space do not follow easily predictable paths and may include out-of-plane motion and non-circular, aperiodic behavior that is difficult to succinctly describe and visualize (see Figure 62)."

AI-95 """The eflightects of the COVID-19 pandemic continue to impact space-related supply chains. The supplier base is not ready for rapid expansion and growth into Cislunar space. If government and private investors rapidly increased funding into the space industrial base for Cislunar development, it would take several years before the supplier base could respond to that increase in demand. This is primarily due to limits on how quickly the space industry can ramp up highly-specialized hiring, facilities, equipment, design and integration capabilities. New technologies, especially those related to space, require sustained investment, policy updates, and focus over years to be successful."""

AI-96 "Despite these challenges, the space industrial base is in a good position for a slow growth proflightle. Unfortunately, slow growth into Cislunar space may not be fast enough to maintain U.S. dominance in this domain."

AI-97 "In 2001, China publicly announced its specifirstc plans to send missions to the Moon under the Chang'e program. This ambitious program included developing communications infrastructure, robotic missions to the far side of the Moon, sample returns, and eventually human missions. These missions have implicit and explicit dual-use military applications. Early U.S. attention to this plan and its implications would have driven us to realize the need for (at a minimum) a Cislunar Space Domain Awareness (SDA) mission. A small investment a decade ago would have led to mature capabilities today." - USSF SPACE FUTURES WORKSHOP REPORT, 2021

AI-98 ""The universe is an ocean, the moon is the Diaoyu Islands [Senkaku Islands, East China Sea], Mars is Huangyan Island [Scarborough Shoal, South China Sea]. If we don't go there now even though we're capable of doing so, then we will be blamed by our descendants. If others go there, then they will take over, and you won't be able to go even if you want to. This is reason enough."

- YE PEIJIAN, Head of China's Lunar Exploration Program, 2018"

AI-99 "A Robust U.S. Industrial Base is a National Imperative - Participants share a common theory of 'causing national security and building enduring military advantage': The secret sauce of America's strategic power has been the conscious and deliberate nurturance of strategic industries for seafaring and aviation. Strategic strength comes from economic and industrial strength, enabled by the transportation modes, which give access to a strategic domain and manufacturing capability to field."

AI-100 "Using what Albert Einstein called the 'magic of compound interest,' the DOD can make a small one-time investment to catalyze a self-sustaining and scalable industry, which it can draw upon sometime in the uncertain future."

AI-101 n/a

AI-102 n/a

AI-103 n/a

AI-104 n/a

AI-105 "A long-term, national space strategy integrating civil, commercial and national security space lines of effort must be developed to retain the U.S.' dominant and leadership position in the emerging future of space. This strategy must account for the possible space futures developed in the workshop."

AI-106 "The 2060 space world will be highly complex and diverse as to the number of state and non-state actors, their capabilities, and their interests. Commercial space presents unique issues as to ownership and sovereignty that, if not resolved, could lead to commercial space entities as independent or semi-independent space powers, resulting in significant opportunities and challenges to U.S. space power. Space power will be widely distributed, making it impossible for any one nation or entity to have predominant space power in the civil, commercial, and military domains. The diversity and distribution of space power enables a wide range of alliances, partnerships, and shared interest. These relationships will be diverse and vary with time as the interest and capabilities of space faring entities develop and change. This complexity poses

AI-107 "A robust/competitive U.S. Space Industrial Base is essential to maintaining the U.S. as a preeminent space power, but its competitive advantage is threatened by increasing globalization of space industrial capabilities."

AI-108 n/a

AI-109 "Most companies: Cited unpredictable U.S. Government demand as the greatest hurdle to growth. Cited export controls and ITAR-restrictions in their top-4 hurdles to growth."

AI-110 This indicates that the Space Industrial Base (SIB) is predominantly small, early stage businesses aligned with demand signals from the government and private industry. Most have tremendous potential for growth in a globally competitive new space market.

AI-111 "The uncertainty of the global pandemic weighed down many companies in early 2020. Venture capital and the public markets have responded positively by investing more dollars into commercial space companies over the past year than in any year prior. It is worth noting, however, that a majority of today's most successful commercial space startups have government business in the form of prototype, procurement and service contracts. Investors took note of this fact and observed that the non-dilutive government money did not 'dry up.'"

AI-112 "However, respondents noted the difficulty of actually applying and getting a SBIR to be high, with many calling the process 'difficult', 'confusing', and 'not user-friendly.' Companies noted a 'valley of death' beyond Phase 2, due to the difficulty in connecting their technology to a program of record, stalling their traction with the DOD."

AI-113 "A quarter (26%) of respondents are considering SPACs as a viable exit strategy for their companies. All of these respondents are non-traditional vendors. The primary cited reason for considering a SPAC is immediate access to capital, along with freedom from recurring fundraising activities."

AI-114 "The shock is unlikely to resolve quickly. Supply uncertainty will beget artificial demand spikes as companies seek to build inventory to buffer against future shortages, like shoppers buying more toilet-paper than usual at the sight of empty shelves. Adding capacity takes years and,

AI-114 (cont.) "while necessary, won't come in time to address the current state. Some project significant shortages are likely through the beginning of 2023."

AI-115 "The government's demand signal was deemed as by 42 of the 57 survey respondents (~74%) which was seen as a hurdle to future growth. Speciflightc challenges included access to key points of contact in government agencies and/or the responsiveness of those POCs, as well as the need for a security clearance and/or an existing contract in order to conduct substantial business development activities. Others felt the government needed to make more long-term commitments to support long-term fundraising and investment. Additionally, many vendors expressed that the contracting process remains lengthy and cumbersome. Some recommended leveraging fully-negotiable Other Transaction Agreements, such as DIU's Commercial Solutions Opening (CSO), to get vendors under contract more rapidly and flightexibly (i.e.

AI-116 n/a

AI-117 "Rocket Science Rut - Stubbornly high launch costs, largely unchanged for over 50 years, have been the primary barrier to the expansion of the space industry."

AI-118 "To remain relevant in the rapidly transforming space industry, the government must move away from its traditional acquisition approach and adopt a more venture-like attitude toward investment and acquisition. This transition could entail suffering some outright failures to achieve one spectacular home run."

AI-119 "The Hybrid Space Architecture is the integration of emergent 'new space' smallsat capabilities with traditional U.S. Government space systems. This evolving resilient architecture will use a 'variable trust' network framework for rapid and secure data exchange among proliferated satellite systems and services that are large and small; government and commercial; U.S. and Allied; in various, diverse, and layered orbits. The architecture shifts from a platform-centric to an information-centric paradigm."

STATE OF THE SPACE INDUSTRIAL BASE 2021
Infrastructure & Services for Economic Growth & National Security

Summary Report by:

J. OLSON,[1] S. BUTOW,[2] E. FELT,[3] T. COOLEY,[3] & J. MOZER[1]
[1]United States Space Force, [2]Defense Innovation Unit and [3]Air Force Research Laboratory

Edited By:
PETER GARRETSON

November 2021

DISTRIBUTION STATEMENT A. Approved for public release: distribution unlimited.

DISCLAIMER

The views expressed in this report reflect those of the workshop attendees, and do not necessarily reflect the official policy or position of the U.S. Government, the National Aeronautics and Space Administration (NASA), the Department of Defense, the U.S. Air Force, or the U.S. Space Force. Use of NASA photos in this report does not state or imply the endorsement by NASA or by any NASA employee of a commercial product, service, or activity.

Cover: Illustration depicting a modular and serviceable space outpost or port solutions (Source: Arkysis)

November 2021

ABOUT THE AUTHORS

Brigadier General John M. Olson, USAF
Mobilization Assistant to the Chief of Space Operations, USSF

Steven J. Butow
Space Portfolio Director, Defense Innovation Unit, OUSD R&E

Colonel Eric Felt, USSF
Director of the Air Force Research Laboratory's Space Vehicles Directorate

Dr. Thomas Cooley
Chief Scientist of the Air Force Research Laboratory's Space Vehicles Directorate

Dr. Joel B. Mozer
Director Science, Technology and Research, USSF

ACKNOWLEDGEMENTS FROM THE EDITOR
Peter Garretson

The authors wish to express their deep gratitude and appreciation to NewSpace New Mexico for hosting the State of the Space Industrial Base 2021 Workshop in Albuquerque, NM; and to all the attendees, whether live or virtual, who spent the time and resources to share their observations and insights to each of the five working groups. The workshop and this report would not have been possible without the dedicated efforts of the working group chairs and co-chairs: Gordon Roesler, Venke Sankaran, Karl Stolleis, AJ Metcalf, Steve Nixon, Payam Banazadeh, Rex Riddenoure, Dennis Poulos, Mandy Vaughn, Katherine Koleski, nor without the outstanding contributions of our guest speakers Lt Governor Howie Morales, Casey DeRaad, Jessica McBroom, Dr. Bhavya Lal, Jay Santee, Bruce Cahan, Sean Ross, Mandy Vaughn, Steve Nixon, Dr. Mir Sadat, Gordon Roesler, Chris Quilty, Bill Woolf, Mark Massa, Julia Siegel, and Clementine Starling.

The virtual workshop would not have been possible without the incredible support provided by Casey DeRaad, Scott Maethner, Arial DeHerrera, Lauren Rogers, David Ryan, Rogan Shimmin, Ryan Weed, Russel Stanton and Klay Bendle. We also wish to thank David Martin, Johanna Spangenberg Jones, Alexandra Sander, Ritwik Gupta and Ric Mommer for their finishing touches.

ABOUT THE KEY GOVERNMENT CONTRIBUTORS

U.S. Space Force | spaceforce.mil

The U.S. Space Force (USSF) is a military service that organizes, trains, and equips space forces in order to protect U.S. and allied interests in space and to provide space capabilities to the joint force. USSF responsibilities include developing military space professionals, acquiring military space systems, maturing the military doctrine for space power, and organizing space forces to present to our Combatant Commands.

Air Force Research Laboratory | afresearchlab.com/technology/space-vehicles/

The Air Force Research Laboratory's mission is leading the discovery, development, and integration of warfighting technologies for our air, space and cyberspace forces. With its headquarters at Kirtland Air Force Base, N.M., the Space Vehicles Directorate serves as the Air Force's "Center of Excellence" for space research and development. The Directorate develops and transitions space technologies for more effective, more affordable warfighter missions.

Defense Innovation Unit | diu.mil

The Defense Innovation Unit's (DIU) mission is to accelerate commercial innovation for national security. It does so by increasing the adoption of commercial technology throughout the military and growing the national security innovation base. DIU's Space Portfolio facilitates the Department of Defense's ability to access and leverage the growing commercial investment in new space to address existing capability gaps, improve decision making, enable a shared common operating picture with allies, and help preserve the United States' superiority in space.

DISTRIBUTION STATEMENT A. Approved for public release: distribution unlimited.

This page was intentionally left blank.

TABLE OF CONTENTS

FOREWORD	v
EXECUTIVE SUMMARY	1
INTRODUCTION	3
A NATIONAL NORTH STAR VISION FOR SPACE	37
SPACE MOBILITY & LOGISTICS	53
SPACE POLICY & FINANCE TOOLS	61
SPACE INFORMATION SERVICES AND THE HYBRID SPACE ARCHITECTURE	71
EVOLVING SPACE OPERATIONAL AND SUPPORT CONCEPTS	79
PERVASIVE SPACE TECH & SUPPLY CHAIN ENABLERS BEYOND LOW EARTH ORBIT	85
EPILOGUE	91
APPENDIX A - WORKSHOP PARTICIPANTS	A-1
APPENDIX B - PREVIOUS REPORTS & KEY RECOMMENDATIONS	B-1
APPENDIX C - 2021 DIU SPACE PORTFOLIO COMPANY SURVEY	C-1
APPENDIX D - INDEPENDENT ANALYSIS OF SPACE INDUSTRIAL BASE	D-1
APPENDIX E - HYBRID SPACE ARCHITECTURE	E-1
APPENDIX F - ACRONYMS & ABBREVIATIONS	F-1

> *"The United States of America has no intention of finishing second in space. This effort is expensive — but it pays its way for freedom and for America."*
>
> – PRESIDENT JOHN F. KENNEDY, 1963[1]

Illustration © James Vaughan and used by permission.
More of his work can be found here: http://www.jamesvaughanphoto.com/

---

[1] Remarks prepared for delivery at the Trade Mart in Dallas, TX, November 22nd, 1963 [Undelivered] as JFK was assassinated while en route to the event.

FOREWORD

The State of the Space Industrial Base 2021 highlights the opportunities and challenges for the United States in space. This report captures key information and includes recommendations to advance the economic, scientific, technological and national security of the nation and our partners; prevail in strategic competition; bolster a US-led international order and enable standards for space; mitigate climate challenges; and boldly expand U.S. leadership in space exploration and discovery to benefit all of humanity and become a true spacefaring nation. While these recommendations do not represent the official position of the United States Space Force, NASA or any other branch of government, they are extremely valuable inputs for consideration.

The 2020 United States National Space Policy directs action to "strengthen and secure the United States space industrial base." To meet that mission, we must foster the security and resilience of the domestic space industrial base and promote the availability of space-related industrial capabilities in support of national critical functions. It also requires us to consider how we strengthen the security, integrity and reliability of supply chains; incentivize key science, technology and industrial suppliers to remain in the United States; seek to create educational and professional development opportunities for the space workforce; and support innovative entrepreneurial space companies through deregulatory actions.

Both NASA and the Space Force believe strong engagement from across government, industry and academia is essential to meet this moment. We applaud this effort to supply thoughtful and actionable recommendations from the very entrepreneurs and innovators who will propel and sustain this strategic American imperative, thereby allowing us to push out further into the cosmos.

Together, NASA and the Space Force seek to advance, protect, and sustain activities in, from, and to space. We do this because it is a vital scientific and economic sector in a manner consistent with the democratic values shared by our nation's allies, friends and regional partners. This is truly a new era of strategic collaboration that will benefit commercial, civil and national security in space.

Just as important, this report provides constructive input for an emerging United States vision for space. It will undoubtedly inspire Americans, providing unity around our spacefaring future to directly address our national priorities and continued exploration of the heavens. To do this, we must: guide the framing of policy and plans to accelerate the development of a vibrant in-space and from-space economy; establish best practices, norms, and interoperability standards; accelerate the build-out of a more robust Hybrid Space Architecture for space science, exploration and information services; create a bold new 21st century space transportation and logistics infrastructure; and mobilize the space sector to address energy demand and climate change.

These insights and recommendations have the power to unite and unleash the full innovative technological, entrepreneurial, and industrial capacity of the United States. Our vision is a bright and bountiful space future that is safe, secure, sustainable, synergistic, and successful.

SENATOR BILL NELSON
NASA Administrator

GENERAL JOHN W. RAYMOND
Chief of Space Operations, United States Space Force

EXECUTIVE SUMMARY

Figure 1: "*The Next Giant Step*" (Artist: Robert McCall, Credit: NASA)

Central Message - This report paints a picture of a U.S. space industrial base that is ***tactically strong but strategically fragile***. While the pace of innovation and investment in the U.S. is at an all-time high, participants cautioned that this *will not be sustained without strategic direction, robust adoption of commercial space capabilities expressed in meaningful contract opportunities, strategic workforce development, attention to fragile domestic supply lines*, and addressing the *anemic funding* to prototype, validate and accelerate the adoption of innovative and disruptive space capabilities for national security.

Major Themes - Major themes articulated by participants included how space is relevant to and must be framed in relationship to America's priorities of infrastructure, 'building back better' our economy with 21st century manufacturing, infrastructure[2] and jobs,[3] tackling climate change, unlocking the full diversity of America's talent, and serving as a platform for foreign policy[4] soft-power and public diplomacy.

Major Opportunities - Participants asserted major opportunities exist to set a national vision; to build an inclusive Cislunar economy; to set key technical and behavioral standards which advantage democracies; to leverage existing commercial satellite capabilities for a Hybrid Space Architecture enabling Joint All Domain Command and Control (JADC2); to create a space internet; to build a space superhighway for logistics; and to apply space solutions to Presidential[5] and Secretary of Defense[6] priorities for climate change.

---

[2] White House (2021). FACT SHEET: President Biden Announces Support for the Bipartisan Infrastructure Framework.
[3] Biden, J.R. (2020). Why America must lead again: Rescuing U.S. foreign policy after Trump. Foreign Affairs; White House (2021). Remarks by President Biden on America's Place in the World.
[4] White House (2021). Interim National Security Strategic Guidance.
[5] White House (2021). Executive Order on Tackling the Climate Crisis at Home and Abroad.
[6] U.S. DOD (2021). Statement by Secretary of Defense Lloyd J. Austin III on Tackling the Climate Crisis at Home and Abroad.

Major Concerns Requiring Urgent Action - However, the participants cautioned that *to secure American leadership, numerous near-term issues must be addressed* to sustain the current pace, including providing strategic guidance, providing adequate resourcing to accelerate innovation, attention to brittle supply chains, and awarding of procurement and services contracts significant enough to sustain current private investment levels. (These recommendations do not represent a USG official position.)

> *"A robust/competitive U.S. Space Industrial Base is essential to maintaining the U.S. as a preeminent space power, but its competitive advantage is threatened by increasing globalization of space industrial capabilities. Other nations are aggressively expanding their space industrial bases. To meet these challenges, the USSF must lead and develop an all-of-government strategy to partner with the U.S. Space Industrial Base, taking advantage of present, commercial capabilities while stimulating future commercial capability development for U.S. space military needs."*
>
> – USSF SPACE FUTURES WORKSHOP REPORT, 2021

Major Recommendations

Attendee recommendations for the White House & Space Council

1. Establish "Space Development and Settlement" as our National "North Star" Space Vision
2. Build Back Beyond: Incorporate the Moon into the Earth's Economic Sphere by Catalyzing the Space Superhighway
3. Sustain funding for the Hybrid Space Architecture as a foundation for the future Space Internet
4. Expand "Artemis Accords" Beyond NASA
5. Increase Space Science & Technology Funding to Parity with Other Domains
6. Reform Policy to Address 21st Century Conditions
7. Declare Space a Special Economic Zone and Deploy the Full Range of Tools
8. Recognize Space-critical Infrastructure / Make Space a Part of Infrastructure Plans
9. Make Space a Central Part of Climate Action Plans
10. Include Space in Supply Chain Planning

Attendee recommendations for the DOD

11. Integrate JADC2 with the Hybrid Space Architecture
12. Enable the Space Superhighway by Including Commercial Solutions for In-space Logistics Infrastructure
13. Mandate a Percentage of Commercial Services Buys Starting in 2022
14. Expand Use and Management of Space Commercial Services within the Space Force
15. Bolder Acquisition Reform Means a More Level Playing Field for All Business, Particularly Small Business
16. Enable Rapid Innovation by Shifting Resources from SBIRs to OTAs

Attendee recommendations for Venture Capital and Investors

17. Balanced Growth Requires Investment Beyond LEO
18. Expand Investments in Enabling Technologies

INTRODUCTION

"Leadership in space exploration has a real political meaning...Failure in that leadership means inevitably falling into the status of a second-class nation with the heavy costs to our way of free enterprise which subjugation to others would involve." – LLOYD V. BERKNER, 1960[7]

This report represents the collective voice of 232 industry experts[8] who gathered to provide inputs and recommendations to nurture and grow a healthy space industrial base and national security innovation base. While these recommendations do not represent the official position of the United States Space Force, or any other branch of government, they are extremely valuable inputs for consideration. The intended audience includes the Administration, National Space Council, senior policymakers across the executive departments, Congress, the U.S. Venture Capital (VC) and investor community, and the broader commercial space ecosystem. The main body of the report provides an overall assessment and general recommendations. It is followed by more in-depth assessments of the current state, challenges, inflections and recommendations needed to preserve America's leadership, as explored by the five workshop teams representing areas of most significant commercial, civil and national security space activity: information services, space logistics, operational concepts, enablers, policy and finance.

GENERAL OBSERVATIONS

Central Message - We assess the U.S. space industrial base as ***tactically strong but strategically fragile***. While the pace of innovation and investment in the U.S. is at an all-time high, participants cautioned that this *will not be sustained without national strategic direction, meaningful commercial contract opportunities with defense and intelligence agencies, attention to fragile domestic supply lines*, and addressing the *anemic research, development, test and evaluation funding* to prototype, validate and accelerate the adoption of innovative and disruptive capabilities required to retain U.S. leadership in space.

How did we get here? First, by failing to recognize the rising threat of China - a nation focused on displacing the U.S. as the predominant space power by 2049; and second, by failing to employ all instruments of national power to accelerate and synergize advancements in commercial, civil and national security space (see Figure 2). The situation is correctable, but it is a fleeting opportunity as the U.S. attempts to outpace and compete with its greatest threat to national security in the first half of the 21st century.

This report provides an American path forward facilitated by a national 'North Star' vision for space.

A Mixed Scorecard - As reported below, while certain commercial aspects of the industrial base are strong and strengthening (e.g. investment, growth, innovation, technology, entrepreneurship, atmospherics, markets and accessibility), more needs to be done to highlight space as a national priority, to make it a part of national strategy, and to reduce risks and provide incentives. Fragility in

---

[7] Wingo, D. (2015). The early space age: the path not taken but now? (Part 1). Retrieved from wordpress.com
[8] See Appendix A.

cybersecurity and national supply chain create risk. New initiatives such as the Hybrid Space Architecture are designed to mitigate some of these risks. More needs to be done to create synergies between different sectors, with alliance partners, and to engage more diverse networks of human capital, including universities. The current momentum of 'strong and strengthening' will not continue without addressing these fragilities.

Figure 2. U.S. Space Industrial Base Scorecard.[9]

Major Themes - Major themes articulated by the participants included how space is relevant to and must be framed in relationship to America's priorities of infrastructure, 'building back better' our economy with 21st century manufacturing, infrastructure[10] and jobs[11], tackling climate change, unlocking the full diversity of America's talent, and serving as a platform for foreign policy[12] soft-power and public diplomacy.

Major Opportunities - Participants asserted that major opportunities exist to set a national vision, to build an inclusive Cislunar economy, to set key technical and behavioral standards which advantage democracies, to leverage existing U.S. and allied commercial space capabilities for a hybrid network architecture enabling the Joint All Domain Command and Control (JADC2), a space internet, a space superhighway for logistics, and to apply space solutions to Presidential[13] and Secretary of Defense[14] priorities for climate change.

Major Concerns Requiring Urgent Action - However, the participants cautioned that *to secure American leadership, numerous near-term issues must be addressed* to sustain the current pace, including providing strategic guidance, providing adequate resourcing to enable innovation, attention to brittle supply chains, and awarding of defense and intelligence contracts significant enough to sustain current private investment levels.

---

[9] Presentation by Gen Olson, U.S. Space Force acting Chief Technology and Innovation Officer at SSIB'21 plenary.

[10] White House (2021). FACT SHEET: President Biden Announces Support for the Bipartisan Infrastructure Framework.

[11] Biden, J.R. (2020). Why America must lead again: Rescuing U.S. foreign policy after Trump. Foreign Affairs; White House (2021). Remarks by President Biden on America's Place in the World.

[12] White House (2021). Interim National Security Strategic Guidance.

[13] White House (2021). Executive order on tackling the climate crisis at home and abroad.

[14] U.S. DOD (2021). Statement by Secretary of Defense Lloyd J. Austin III on Tackling the Climate Crisis at Home and Abroad.

FRAMING SPACE IN THE PUBLIC DISCOURSE

> *"Whoever is first to establish a sustainable presence on the Lunar surface will retain both the perception and reality of national preeminence."* – DR. BHAVYA LAL, NASA, 2021

Space is Key to Winning the Future - The participants told us it is time for the dominant policy image of space to change, and that a new public discourse is needed to frame space as an economic domain, as a major source of 21st century industry and jobs, as a source for climate change solutions, and as a platform for soft-power and knitting together democracies. The participants advanced a number of propositions for how the Administration, DOD and Congress should view and frame space:

Space is an Economic Domain - While much has been said about space as a domain of science and discovery and while space is now acknowledged as a warfighting domain, too little is said about space as an economic domain[15] -- an economic domain of vast opportunity, and a domain of citizen and private activity and commerce. Commerce, however, is the principal contribution of space to U.S. power, influence and prosperity.

Space is a Strategic Domain - Space has become central to strategic competition and grand strategy; the challenge to our national prestige, leadership and grand strategy is clearly being made by strategic competitors in space. As one participant remarked, "First in space. First in everything. Second in space, second in everything."[16]

Figure 3: Space is vital to U.S. economic and national security.[17]

Space is Critical to U.S. Soft Power - Space is the most public stage to display aspirational audacity, societal vibrancy, technical competence, and measurable achievement. Whoever is first to establish *sustainable* presence on the Lunar surface will retain both the perception and reality of national preeminence, and be able to leverage substantial soft power to achieve geo-political objectives and support their nation's values, and could lead to a de facto leadership role in defining broad principles of

---

[15] Remarks by Gen Olson, U.S. Space Force acting Chief Technology and Innovation Officer at SSIB'21 plenary.
[16] These words were earlier spoken by Livio Scarsi as described in Ettore Perozzi, Sylvio Ferraz-Mello, Space Manifold Dynamics: Novel Spaceways for Science and Exploration, p. 230
[17] Slides presented by Gen Olson, USSF (see ref 14 above)

behavior for the Moon and Cislunar space.[18] If led by the U.S., it offers opportunities to deepen its alliance with international partners, including possibly Russia, while expanding future collaborations to include new partners in the Middle East, Asia, Africa, and South America. If this competition is won by China, they will leverage the substantial soft power it would gain from Lunar preeminence to achieve a wide variety of national security, economic, and diplomatic/political objectives. The global perception of the United States will be substantially wounded, feeding a narrative of American decline and China's ascendance. China conducting these activities in partnership with Russia would have even graver implications for national security.[19]

Space is Critical to U.S. Hard Power - Our nation's deterrence and warfighting advantage rests on space capabilities to provide peacetime indications and warning, missile warning, over the horizon communications, overhead targeting, precision timing for coordination, precision navigation, and precision targeting.[20]

Space is a Transportation Mode - [21] Like land, sea, and air, space represents a distinct mode of transportation. Already, American private industry produces commercial vehicles to access low Earth orbit with passengers and cargo up to 66 metric tons, cargo services to MEO and GEO, and this year will demonstrate parcel-class cargo to the Moon. Within five years industry is expected to offer commercial services of 100MT to LEO and both passenger and heavy cargo to GEO, the Moon, and Mars as well as between locations on Earth.

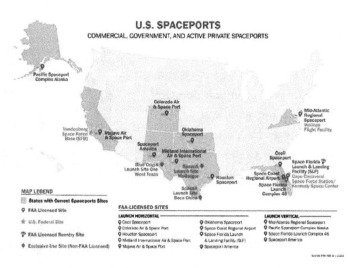

Figure 4: U.S. Spaceports (Source: FAA.gov).

Space deserves to be elevated to peer status with other modes in the Department of Transportation, and space ports deserve treatment and federal improvement funding similar to airports.[22]

---

[18] Duffy, L. and Lake, J. (2021). Cislunar Spacepower, The New Frontier. Space Force Journal.
[19] Remarks by Dr. Bhaya Lal, Senior Advisor to the NASA Administrator for Budget and Finance at SSIB'21 plenary.
[20] Remarks by Gen Olson, U.S. Space Force acting Chief Technology and Innovation Officer at SSIB'21 plenary.
[21] Presentation by General Steve "Bucky" Butow at SIB 2021 plenary
[22] Messier, D. (2021). FAA Limits Evaluation of Spaceport Infrastructure Funding Options. Parabolic Arc.

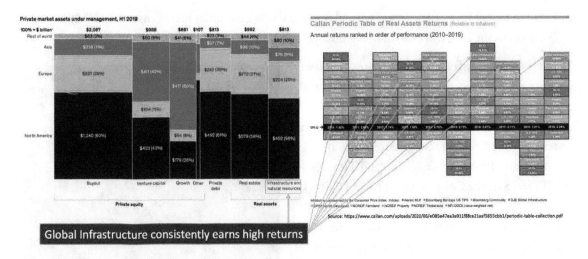

Figure 5: Investors seek global infrastructure's high returns.[23]

Space is Infrastructure - The assets we have in space are best thought of as infrastructure. Space assets underpin nearly every sector of our society. They synchronize our power grid. They synchronize, coordinate and secure our financial transactions. They connect our cities and rural areas, providing long-distance communications for television, radio, telephony and broadband internet. They supply weather, traffic and logistics data to enable city planning, agriculture, public health and transportation. They provide advance warning of a broad range of hazards including extreme weather events to safeguard lives, crops, and critical infrastructure. Conceptualizing space assets as infrastructure can open new financial tools, and broaden the diversity of investors who can participate in the space economy beyond just venture capital investors. As noted in *U.S. Space Policies for the New Space Age: Competing on the Final Economic Frontier*, treating mature space assets as an infrastructure class of investments enables space companies to raise capital via traditional infrastructure instruments such as space bonds.[24] Such financial instruments open the space economy to a wider pool of "space" investors

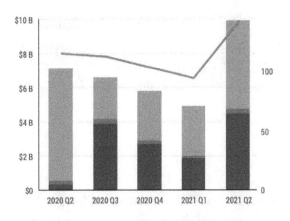

Figure 6: Q2 2021 was the fourth largest quarter on record for investment in the space economy, and the largest quarterly investment in space infrastructure (dark green) ever recorded.[25]

which include longer term, private equity and institutional investors such as insurance companies, pension funds, and university endowments. Institutional investors often seek global infrastructure investments because they consistently earn high returns. Offering space as an alternative infrastructure investment unlocks a wider pool of additional investors. This pool is substantial, as investments

---

[23] With permission from Bruce Cahan and Urban Logic Inc. (2017).
[24] Cahan, B. and Sadat, M. (2020). U.S. Space Policies for the New Space Age: Competing on the Final Economic Frontier. pages 73 - 75
[25] Space Capital (2021). Space Investment Quarterly Q2 2021. Space Capital.

allocated to global infrastructure and natural resources as "assets under management" (AUM) exceed $473 billion in the U.S. and $880 billion globally.[26]

Space is Critical Infrastructure - Our space assets are not only infrastructure, they are infrastructure critical to the functioning of our economy and society.[27] Our financial system, telecommunications system, power water and sewage systems are all dependent on Precision Navigation and Timing (PNT) and Satellite Communication (SATCOM). The loss of GPS service would average a $1 billion per-day impact to the nation.[28] If space is critical infrastructure, impatient venture capital alone is not enough to compete with the pacing challenge.[29] The Cybersecurity and Infrastructure Security Agency (CISA) established a Space Systems Critical Infrastructure Working Group in May 2021 to identify and develop strategies to minimize risks to space systems that support the nation's critical infrastructure.[30]

Figure 7: The fifth modernized GPS III satellite reached orbit on 17 June 2021 (Lockheed Martin).

There are sixteen critical infrastructure sectors whose assets, systems and networks are considered so vital to the United States that their incapacitation or destruction would have a debilitating effect on security, national economic security, national public health or safety, or any combination thereof. Space infrastructure should be added to this list.

> "90% of what we know about climate change, we have learned from space."
> — DR. GORDON ROESLER, 2021

Space is Key to Tackling Climate Change - NASA satellites have made an outsized contribution to our knowledge of climate change.[31] In fact, it is only through space that we even became conscious of climate and its changes — it was efforts to model other planetary surfaces and understand heliophysics[32] which led to the first climate models. Moreover, because the majority of the Earth's surface are oceans,

---

[26] McKinsey & Company (2021). A year of disruption in the private markets: McKinsey Global Private Markets Review 2021. Page 10.
[27] Remarks by Dr. Mir Sadat, former NSC Space at SSIB'21 plenary; Swallow, E. and Visner, S. (2021). It's time to declare space systems as critical infrastructure. Politico.
[28] RTI International (2019). Economic Benefits of the Global Positioning System (GPS): Final Report.
[29] Presentation by Bruce Cahan, Urban Logic, Inc. at SSIB'21.
[30] CISA (2021). CISA launches a space systems critical infrastructure working group.
[31] NASA (2021). Taking a global perspective on Earth's climate.
[32] Frank, A. (2018). Light of the Stars: Alien Worlds and the Fate of the Earth. WW Norton & Company.

space-based observations are critical to providing data in sparse regions.[33] While the perspective offered by Space of Earth is broadly acknowledged[34] to have started the global environmental movement, few policy makers think of space itself as a green technology. Few policy makers seem aware that the improved route planning made possible by GPS reduces global transportation emissions between 15 and 21%.[35] Space technology offers big ideas for climate solutions in six distinct categories:

- Superior Climate Situational Awareness
- Compliance Monitoring
- Materials to Support the Green Economy
- Avoiding Industrial Impact
- Controllable and Reversible Emergency Climate Interventions
- Vast Reservoirs of Green Energy

Some of these are being actively pursued by America's strategic competitors for profit, prestige and to attract international partners. More detail is provided in the Section of this report entitled North Star Vision.

Figure 8: AFRL's Space Solar Power Incremental Demonstrations and Research Project (SSPIDR) is a collection of flight experiments designed to mature critical technologies needed to build an operational solar power beaming system in space transforming energy production & distribution on Earth (AFRL).

Space is a 21st Century Industry - In the next decade, the U.S. has the means to lay the foundations both for in-space resource extraction (Lunar and asteroid mining) and in-space manufacture as well as the supporting transportation and logistics. The space economy is poised to expand from an economy built solely around 'bits,' to an economy delivering "bits joules and atoms"[36] or data, energy and raw materials. It has the capacity to rapidly advance autonomy, artificial intelligence, and robotics. It will enable low gravity and vacuum manufacturing. This may also include biotechnology, such as in-space 3D printing of retinas[37] and hearts[38] and other organs.[39] Space may become a destination for voluntary travel[40] with a supporting hospitality industry. Emerging space based information services such as commercial broadband, remote sensing (including live video,[41] hyperspectral, radar, radio frequency (RF) mapping), and emerging ultra-rapid suborbital point-to-point transportation can empower even more industries on Earth, including precision terrestrial logistics, supply chain management, safer

---

[33] Jiménez Alonso, E. (2018). Earth observation of increasing importance for climate change adaptation. Acclimatise.
[34] Frank, A. (2018). *Light of the Stars: Alien Worlds and the Fate of the Earth*. WW Norton & Company.
[35] Autry, G. (2019). Space research can save the planet - again. Foreign Policy.
[36] Bloxton, M. (2020). Bits Atoms Joules. Medium.
[37] DeBos, C. (2020). LambdaVision wants to produce artificial retinas in space. The Burn-In.
[38] Techshot (2020). Success: 3D Bioprinter in Space Prints With Human Heart Cells. Cision PRNewswire.
[39] Sims, J. (2021). Why astronauts are printing organs in space. BBC.
[40] Several companies have announced plans for private space stations and 'space hotels'.
[41] Examples include Canada's UrtheCast and the UK's Sen; and O'Callaghan, J. (2020). This Space TV Startup Plans To Stream Live Videos Of Earth's Surface From Space In 2021. Forbes.

transportation, urban planning, and precision agriculture. All of this is possible, unless the U.S. elects to cede its leadership role in space to China through inaction or atrophy.

Figure 9: Starship being mated with its Super Heavy Stack (Credit: SpaceX)

> *"Financing [Space Solar Power Satellites] a little-known element of NASA's union-built, clean energy technology through that infrastructure bill is the fastest, most painless way to accomplish, and pay for, the Green New Deal (GND) and President Biden's goal to Build Back Better... It will also bring unexpected "space benefits" to Native American, Black and Latino populations across the nation."*
> – INT'L ASSOCIATION OF MACHINISTS AND AEROSPACE WORKERS, 2021[42]

Space is a Source of New Jobs - Space has always been a source of jobs and a creator of Science Technology Engineering and Math (STEM) talent. But the future could hold much more. Current projections suggest that today's space sector, which employs approximately 420,000 and generates approximately $210 billion is expected to grow to $780 billion and 1.5 million jobs by 2050 under current incentives.[43] But, according to the Foundation for the Future, *just unlocking traditional infrastructure development financial tools* for space could mean *an additional 195,098 jobs and additional $20 billion of gross space product by 2025*, and an astounding *3,107,318 additional jobs and a $445 billion larger space sector* than the current approach by 2050.[44] Moreover, if the U.S. space industry leads in space-based green energy and succeeds in developing grid competitive power, the International Academy of Astronautics (IAA) estimated that this new service would *eventually grow to five million new jobs*, equivalent in scale to the hospital or fast food industries in numbers employed.[45]

---

[42] Myers, E. (2021). *How President Biden, Congressional Democrats, NASA and the 2021 Infrastructure Bill Can Unite America and Achieve the Job Creation, Climate Change and Social Justice Goals of the Green New Deal.* International Association of Machinists and Aerospace Workers
[43] Foundation for the Future (2021). U.S. space economy through 2050.
[44] Ibid. (43)
[45] Mankins, J. (2011). Space Solar Power. IAA.

That includes skilled labor. While most 'green jobs' average $20/hr, aerospace pay rates exceed $30/hr, even $50/hr including benefits, and aerospace workers spend more in their community.[46]

Space is America's Competitive Advantage - Aerospace has always been America's strongest sector, leading exports. We have both the strongest tech base, most desirable environment for intellectual property and the most secure and well capitalized finance environment. The total aerospace industry generated $148 billion worth of exports in 2019, provided a positive trade balance of over $79 billion[47] and employs a workforce of 2.4 million people.[48] The advent of commercial suborbital point-to-point provides yet another market segment where proactive U.S. policy could sustain a natural U.S. advantage where U.S. manufacturers and carriers are the global flag of choice.

Space is a Source of Economic Growth - The global space industry is estimated by the Space Foundation to have been $446.9 billion in 2020. This $447 billion economy is 55% larger than a decade ago, and part of a five-year trend of uninterrupted growth.[49] Projections for 2040 range from 1-1.5 trillion (Goldman Sachs $1.1 trillion,[50] Morgan Stanley $1.1 trillion,[51] U.S. Chamber of Commerce, $1.5 trillion[52]), and projections for 2050 range from $2.7 trillion (Bank of America Merrill Lynch[53]) to $10 trillion (China Academy of Space Technology[54]). A total of $177.7 billion of new capital investment has been added to the commercial space industry in the last 10 years and $25.6 billion of that was added just in the last year.[55] Recent estimates also suggest that space tourism could be a $4 billion industry[56] and suborbital point-to-point could be a $20 billion industry by 2030.[57]

Space is a Source of Technological Innovation - Space technology offers improvements in safety and lifestyle on par with other 21st century technologies such as artificial intelligence, biotechnology, green energy, electric vehicles — and in many cases is leading technological development in those. An Earth-independent supply chain can offer energy and strategic materials critical to a green economy without environmental impact to Earth.

Space is an Investment, Not a Must-Pay Bill - America invests its tax dollars in space because it provides a substantial return on investment. Infrastructure such as the USSF-operated Global Positioning System (GPS) has generated $1.4 trillion for the U.S. economy since made publicly available

---

[46] Myers, E. (2021). *How President Biden, Congressional Democrats, NASA and the 2021 Infrastructure Bill Can Unite America and Achieve the Job Creation, Climate Change and Social Justice Goals of the Green New Deal.* International Association of Machinists and Aerospace Workers

[47] AIA (2019). Foreign Trade: Key Trade Statistics. Aerospace Industries Association.

[48] Erwin, S. (2018). Report: U.S. aerospace a trade winner, but tariffs threaten future exports. SpaceNews.

[49] Space Foundation (2021). The Space Report 2021 Q2. Space Foundation.

[50] Morgan Stanley (2020). Space: Investing in the Final Frontier. Morgan Stanley.

[51] Smith, R. (2018). The $1.1 trillion space industry prediction you can't afford to miss. The Motley Fool.

[52] Higginbotham, B. (2018). The Space Economy: An Industry Takes Off. U.S. Chamber of Commerce.

[53] Sheetz, M. (2017). The space industry will be worth nearly $3 trillion in 30 years, Bank of America predicts. CNBC.

[54] Siqi, C. (2019). China mulls $10 trillion Earth-moon economic zone. Global Times.

[55] Space Capital (2021). Space Investment Quarterly Q4 2020. Space Capital.

[56] Miao, H. (2021). UBS says space tourism could be a $4 billion market by 2030. Here's how to play it. CNBC.

[57] Sheetz, M. (2019). Super fast travel using outer space could be $20 billion market, disrupting airlines, UBS predicts. CNBC.

in the 1980s[58] and each year, GPS generates $70 billion for the U.S. economy[59] while GPS-guided navigation enabled a 15% to 21% reduction in transportation fuel expenditures and associated emissions.[60] With a mere total budget authority of $21.5 billion in FY 2019, NASA activities in the U.S generated $64.3 billion in economic output[61] -- a return on investment of around 200%.[62] NASA activities also supported 312,630 jobs in the U.S. in FY 2019, nearly 50,000 of which are in scientific research and development disciplines.[63] That R&D pays off, with $14.2 billion of NASA's generated economic output in the scientific research and development sector[64] and documented cases of over 2,000 spinoff technologies developed by NASA or developed with NASA assistance.[65] Simply put, space dollars are dollars that result in a larger U.S. economy.

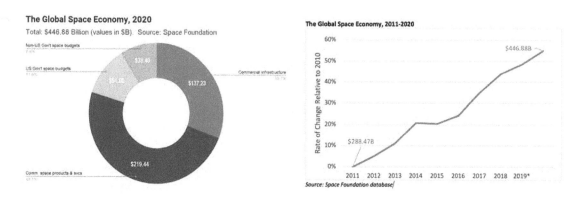

Figure 10: The global space economy was valued at $446.88 billion in 2020, a growth of 55% since 2011.

THE YEAR (2021) IN REVIEW

Since the publication of the State of the Space Industrial Report 2020, enormous changes have taken place. Overall, despite the COVID-19 pandemic, the *space sector maintained and gained momentum.*

Continuity and Change in U.S. Space Policy - In the government sector, a new Administration took the reins establishing new priorities to combat COVID, 'build back better' the economy, tackle climate change, demolish structural barriers to unlock the full potential of America's diverse talent, and unite democracies against autocracies. Recognizing the centrality of space as a theater of competition between democracy and autocracy and the growing bipartisan consensus on the importance of space, the Administration provided continuity of momentum by continuing the National Space Council

---

[58] U.S. Space Force (2020), GPS celebrates 25th year of operation; McTigue, K. (2019). Economic Benefits of the Global Positioning System to the U.S. Private Sector Study. NIST.
[59] U.S. DOD (2021). Protecting America's Global Positioning System. Retrieved from https://www.defense.gov.
[60] Autry, G. (2019). Space research can save the planet - again. Foreign Policy.
[61] University of Illinois at Chicago Nathalie P. Voorhees Center for Neighborhood and Community Improvement & NASA (2020). National Aeronautics and Space Administration & Moon to Mars Program Economic Impact Study.
[62] Daniel Morgan, D. (2021). NASA Appropriations and Authorizations: A Fact Sheet. Congressional Research Service.
[63] Imperato, A., Garretson, P., & Harrison, R. (2021). U.S. Space Budget Report. AFPC.
[64] Ibid. (63).
[65] NASA (2020). NASA Report Details How Agency Significantly Benefits U.S. Economy.

under the Vice President,[66] continuing the Artemis Program and international partners under the Artemis Accords,[67] and provided the full support of the Administration to the Space Force.[68]

America's Competitors Threw Down the Gauntlet - America's competitors continued their efforts to assume leadership. Just in the interval since the last report — one year —the Communist Party of China (CPC) succeeded in its announced goals to execute a Lunar sample return,[69] achieve Mars orbit, land, and deploy a rover,[70] launch and crew its competitor space station[71] and test suborbital transportation systems.[72] China also experienced a surge in commercial investment[73] following space being designated 'new infrastructure' along with 5G.[74] China opened its space station for international experiments via the United Nations.[75] China and Russia announced a joint Lunar base and released a user's guide for nations interested in partnering.[76] China has expended significant political capital committing to big audacious goals to aid its public diplomacy, soft power, attractiveness as a partner, and international prestige. As part of its soft-power space push, China has also mapped the space contributions to each of the UN development goals.[77]

COVID-19 Effects and Recovery - Proactive measures by the government enabled the space industry to survive COVID, and investment is now surging, having reached $7 billion with new financial innovations such as Special Purpose Acquisition Corporations (SPACs).[78] *But there is a need for this investment to move beyond launch services alone to supply capital into other maturing verticals in the space ecosystem.* For an in-depth discussion, see Appendix D. Designating space systems as critical infrastructure will provide better protection of the U.S. space industrial base from unforeseen challenges into the future.

---

[66] Malik, T. (2021). Vice President Kamala Harris to lead National Space Council under Biden administration. Space.com
[67] https://it.usembassy.gov/biden-is-committed-to-nasas-artemis-program-for-the-moon-and-beyond/
[68] Insinna, V. (2021). With Biden's 'full support,' the Space Force is officially here to stay. Defense News.
[69] Clark, S. (2021). Chinese mission returned nearly 4 pounds of lunar samples. Spaceflightnow.
[70] Webb, S., Allen, R. (2021). On its first try, China's Zhurong rover hit a Mars milestone that took NASA decades. The Conversation.
[71] Amos, J. (2021).China space station: Shenzhou-12 delivers first crew to Tianhe module. BBC.
[72] Jones, A. (2021). China launches secretive suborbital vehicle for reusable space transportation system. Space News.
[73] Patel, N. (2021) China's surging private space industry is out to challenge the U.S.. MIT Technology Review; Waidelich, B. (2021) China's commercial space sector shoots for the stars. East Asia Forum.
[74] Qu, T. (2021) China's new bid to take on Elon Musk's Starlink: a state-owned satellite enterprise. South China Morning Post.
[75] Zheng, W. (2021) Ahead of Chinese space station mission, a call for more collaboration. South China Morning Post.
[76] CNSA & ROSCOSMOS (2021). International Lunar Research Station Guide for Partnership (V1.0). CNSA.
[77] ISU (2021). Alumni Conference 2021. YouTube.
[78] Kemp, K. (2021). Start-ups Aim Beyond Earth. New York Times.

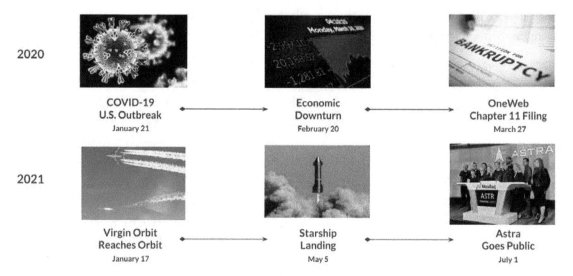

Figure 11: The commercial space sector made an amazing rebound following the unforeseen tightening of the capital markets during the COVID-19 pandemic. The U.S. must be more resilient in the future.

Major Commercial Milestones Were Achieved in 2020-2021 - Both Virgin Galactic and Blue Origin successfully flew their founders, Richard Branson and Jeff Bezos respectively, on high-profile suborbital flights in July 2021,[79] opening the market for suborbital tourism with an already long-list of signed customers.[80] SpaceX successfully launched its first orbital tourist flight, Inspiration4[81] on September 15,[82] and Axiom named the first private space crew for International Space Station tourism, set to launch next year.[83] Jeff Bezos retired from Amazon to focus on Blue Origin.[84] NASA awarded SpaceX the Artemis Human Lander System (HLS) contract in April 2021 to land the first woman and the first person of color on the Moon with a variant of Starship.[85] SpaceX Starship had its first successful launch in August 2020, first successful landing in May 2021,[86] and initial stacking on the SuperHeavy booster in August 2021 in preparation for an expected first orbital launch this year.[87] The Space Force's Rocket Systems Launch Program (RSLP) Office and DOD Space Test Program supported by the DIU continued to prototype commercial launch as a service adding two new providers with launch capacities up to 1,200 kg. A total of 228 DOD payloads were launched by SpaceX on Transporter 1 (Jan 2021),[88] Transporter 2 (Jun 2021)[89] and four additional DOD payloads

---

[79] Wattles, J. (2021). Virgin Galactic founder Richard Branson successfully rockets to outer space. CNN. ; Dunn, M. (2021). Jeff Bezos blasts into space on own rocket: 'Best day ever!'. AP News.
[80] Wattles, J. (2021). Jeff Bezos' Blue Origin to auction 1 ticket for first space tourism flight. ABC11.
[81] Gershgorn, D. (2021). Spacex's Inspiration4 Mission: Launch Date, Crew, Mission Details For The Historic Journey. INVERSE.
[82] Sheetz, M. SpaceX's historic Inspiration4 launch reaches orbit successfully carrying private crew. CNBC.
[83] Gershgorn, D. (2021). Spacex's Inspiration4 Mission: Launch Date, Crew, Mission Details For The Historic Journey. INVERSE.
[84] Mathewson, S. (2021). Blue Origin founder Jeff Bezos will step down as Amazon CEO. Space.com.
[85] NASA. (2021). As Artemis Moves Forward, NASA Picks SpaceX to Land Next Americans on Moon. NASA.gov.
[86] Amos, J. (2021). SpaceX Starship prototype makes clean landing. BBC.
[87] Sheetz, M. (2021). SpaceX aims to launch first orbital Starship flight in July, company president says. CNBC.
[88] Thompson, A. (2021). SpaceX launches a record 143 satellites on one rocket, aces landing. Space.com.
[89] Lentz, D. (2021). SpaceX successfully launches Transporter 2 mission with 88 satellites. NASASpaceflightNow.

were launched by VOX Space (June 2021).[90] These delivered payloads proved useful to solving DOD problems from companies large and small. Capella launched its first commercial SAR satellite, Sequoia, in Aug 2020 and has since followed with four additional Whitney-class satellites to date in 2021. The first Space Development Agency (SDA) satellites were placed on orbit[91] and contracts awarded to SpaceX for its first 150 operational satellites of its tranche 0.[92] Starlink was declared operational in June, and operational globally in 12 countries by August.[93] Northrop Grumman's Mission Extension Vehicle 2 (MEV-2) successfully docked with Intelsat 10-02, paving the way for on-orbit and life extension services.[94] The second quarter of 2021 saw the largest quarterly investment in space infrastructure ever recorded with $4.5 billion of private investment addressing in-space servicing, mobility and logistics.[95] Global space situational awareness is predicted to reach $1.4 billion by 2026, and has already seen ANSYS acquire Analytical Graphics for $700 million,[96] and AFWERX began the Commercial Space Situational Awareness marketplace.[97]

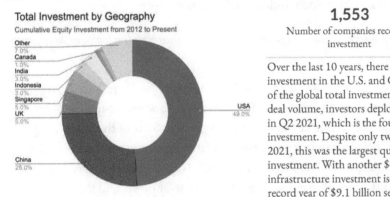

Figure 12: Rise of the Space Startups (Source: Space Capital)

Space is Attracting Substantial Venture Capital - The VC industry continues to be a major contributor to growth of the new space economy. A total of $199.8 billion of new capital investment has been added to the commercial space industry since 2012 and $14.9 billion of that was added in the first half of 2021 alone (see Figure 12). Between 2000-2009, a total of just 49 space-related startups were supported by just 8 participating VC firms. From 2010 to 2018, a much larger 190 startups were funded by a pool of 190 participating VC firms.[98] In just the second quarter of 2021, a total of 138 companies shared a total of $9.9 billion in investment dollars.[99]

---

[90] Strout, N. (2021). Virgin Orbit plane launches four U.S. military satellites into space. Defense News.
[91] Strout, N. (2021). The Space Development Agency now has demo satellites on orbit. Here's what they'll do. C4ISRNET.
[92] Jewett, R. (2021). SpaceX Receives $150 Million Contract to Launch SDA Tranche 0 Satellites. Via Satellite.
[93] Reardon, M. (2021). Elon Musk says Starlink will be available worldwide in August. CNET.
[94] Strout, N. (2021). Virgin Orbit plane launches four U.S. military satellites into space. C4ISRNET.
[95] Space Capital (2021). Space Investment Quarterly Q2 2021. Space Capital.
[96] Strout, N. (2021). Space situational awareness company to be bought for $700 million. C4ISRNET.
[97] U.S. Space Force. (2019). Boosting Space Situational Awareness: SMC Awards SBIR Phase 2 Contract. Spaceforce.mil.
[98] Space Capital (2021). Space Investment Quarterly Q4 2020. Space Capital.
[99] Ibid. (95)

SPACs Have Emerged as a Tool to Raise More Capital - Special Purpose Acquisition Companies (SPAC)[100] raised a record $76.2 billion in 2020, up 557 percent from 2019, and SPAC stocks have delivered average annualized returns of about 17.5 percent since 2015.[101] SPACs allow qualified investors to create a 'blank check' company which goes public with an intention to acquire and consolidate a start-up with high future earnings potential, and merge it into the publicly traded company. This event results in an infusion of capital into the company while providing an exit opportunity to early investors. Ten companies have been announced as commercial SPAC mergers since mid-2019: Astra,[102] AST, Blacksky, Planet, Redwire,[103] RocketLab, Satellogic, Spire, Virgin Galactic, and more recently, Aurvandil.[104] SPACs currently account for $4.8 billion in direct investment in commercial space companies with a total market valuation exceeding $20 billion.[105] This is a major development.

Celebrating Progress on Last Year's Recommendations - Some limited progress has occurred on the recommendations of last year's State of the Space Industrial Base report. We recommended:

1. *U.S. promulgate a "North Star" top-level vision and strategy for space industrial development and establish a Presidential Task Force to execute it.*

    On July 23, 2020, the National Space Council released *A New Vision for Deep Space Exploration and Development*[106] which made some progress toward a whole-of-nation 'North Star' vision for civil space, incorporating key elements highlighted in State of the Space Industrial Base 2020 (SSIB'20). At least one think tank has attempted to provide such a vision. Congress tasked the administration (NSpC) to conduct a U.S.-China Space net assessment and strategy including U.S. equities in Cislunar, space mining, space solar power and launch.[107]

2. *U.S. Department of Defense develops plans to protect, and support and leverage commerce in space.*

    National policy now tasks the Secretary of Defense to "Protect freedom of navigation and preserve lines of communication that are open, safe, and secure in the space domain."[108] Both the USSF[109] and USSPACECOM[110] signed memoranda of understanding with NASA to collaborate on many areas of mutual interest but did not specifically address matters involving the U.S. space industrial base or space commerce. The SSIB'20 report highlighted the

---

[100] Huddleston, T. (2021). What is a SPAC? Explaining one of Wall Street's hottest trends. CNBC.
[101] Bell, D. (2021). SPACs: A hot topic for investors, acquirers and sellers. KPMG.
[102] Foust, J. (2021). Astra to go public through merger with SPAC. SpaceNews.
[103] Foust, J. (2021). Redwire to go public through a SPAC merger. SpaceNews.
[104] Renaissance Capital (2021). Space-focused SPAC Aurvandil Acquisition files for a $250 million IPO.
[105] Multiple sources including SpaceNews, CNBC and other financial news services.
[106] U.S. National Space Council. (2020). A New Era for Deep Space Exploration and Development. Retrieved from Aerospace.org
[107] U.S. Congress (2020). Sect. 1614.Report and strategy on space competition with China, Public Law 116-283, National Defense Authorization Act for Fiscal Year 2021. Govtrack.
[108] White House (2020). The National Space Policy. Federal Register.
[109] NASA (2020). Memorandum of Understanding Between the National Aeronautics and Space Administration and the United States Space Force. NASA.gov.
[110] Smith, M. (2021). USSPACECOM To Sign MOA With NASA Including Cooperation On Planetary Defense. Space Policy Online.

importance of "Clarifying the USSF role in protecting and enabling U.S. commerce across Cislunar space is critical." The USSF has since hosted a Space Futures Workshop to explore and signal this. Toward that end, AFRL has put out a *Cislunar Primer*,[111] and AFRL space leadership articulated this argument in the new *Space Force Journal*.[112]

3. *U.S. Government works to economically stimulate the industry, including space bonds and a Space Commodities Exchange[113] and by executing $1 billion of existing DOD and NASA funding through the Exchange.*

    The DOD consolidated the purchase of its first commodity, commercial SATCOM services under the Commercial Satellite Communications Office (CSCO) of the U.S. Space Force. DOD made progress toward 'launch as a service' with the launch of GPS III on a previously flown Falcon 9 launch vehicle. The procurement of other commercial space services, specifically persistent satellite imagery and data analytics, remains a work in progress.

Figure 13: South Korean Science and ICT Minister Lim Hye-sook displays her signing of the Artemis Accords with the United States (Credit: Ministry of Science and ICT)

4. *U.S. Government develops a framework for creating wealth and security with allies and partners that share our common norms and values.*

    Since publication, the U.S. established a White House policy of encouraging international use of space resources, introduced the Artemis Accords with at least twelve signatories,[114] and has begun G7 discussions about an alternate infrastructure development program to China's Belt and Road Initiative.[115] Former U.S. space officials in think tanks have articulated 30-year strategies for such platforms. Significant diplomatic work remains to promulgate international norms of behavior, operational collaboration, and science and technology partnerships with allies and partners.

---

[111] Holzinger, M., Chow, C. Garretson, P. (2021). A Primer on Cislunar Space. AFRL.
[112] Buehler, E. et al. (2021). Posturing Space Forces for Operations Beyond GEO. The Space Force Journal.
[113] Cahan, B. & Sadat, M. (2021). U.S. Space Policies for the New Space Age: Competing on the Final Economic Frontier. NSNM. pages 76 – 80
[114] NASA (2021). Artemis Accords. NASA: As of June 2021, 12 countries have embraced the Artemis Accords: Australia, Brazil, Canada, Italy, Japan, Luxembourg, New Zealand, the Republic of Korea, Ukraine, the United Arab Emirates, the United Kingdom and the United States.
[115] Wintour, P. (2021). G7 backs Biden infrastructure plan to rival China's belt and road initiative. The Guardian

5. *U.S. Government supplies the workforce necessary to fill more than 10,000 Science Technology Engineering and Math (STEM) jobs domestically.*

   President Biden's FY22 budget request seeks more than $5 billion to expand existing institutional aid grants that can be used to strengthen educational programs for low income and ethnic minorities in high-demand STEM fields.[116] Significant work remains to foster the required work force.

6. *USSF works closely with space industry entrepreneurs and innovators to develop government-commercial technology partnerships that support U.S. commerce and national security in space.*

   Space Force and AFRL have established the SpaceWERX and Rocket Cargo Program, and DIU had successes in prototypes involving small responsive launch, synthetic aperture radar (SAR) smallsats, and space domain awareness (SDA) as subscription services. DIU in partnership with USSF established initial programs to mature in-space development and test of multi-orbit logistics. The SDA has flown its first experiments. USSF established the CSCO. Nevertheless, significant work remains to grow additional commercial technology partnerships.

Figure 14: Rocket cargo enables rapid delivery of aircraft-size payloads for agile global logistics

CURRENT STATE

The United States is well positioned to build on its enduring advantages to meet today and tomorrow's challenges.[117]

Tactically Strong but Strategically Fragile - Despite record levels of private investment in a record number of U.S. space companies offering a steady parade of new products and services, success is hinged on shaky support by the U.S. Government as a consumer of commercial products and services.[118] The notable exception here is NASA, which continues to strengthen both its direct investment in, and procurement of, commercial space offerings. According to participants, a lack of strategic guidance and clarity; the inability to reward private capital investment with meaningful contracts -- to "buy commercial;" a fragile supply chain; and strong international competition and predatory practices by adversaries means that failure to take action puts the gains so far at risk. In April of this year, forty U.S. technology companies co-signed a letter urging the Biden Administration to do more to incentivize the procurement of commercial products and services.[119] The letter specifically requests that the Office of

---

[116] Schwarber, A. (2021). FY22 Budget Request: STEM Education. American Institute of Physics.
[117] The White House (2021). Interim National Security Strategic Guidance. Whitehouse.gov.
[118] Cohen, S. (2018). New Entrants and Small Business Graduation in the Market for Federal Contracts. CSIS.
[119] Censer, M. (2021). Technology companies, associations urge Biden administration to push for more commercial products. Inside Defense.

Management and Budget provide clear guidance to federal agencies to make certain that the existing statutory requirements for commercial preference are followed. Furthermore, nine 4-star generals co-signed a memo addressed to the intelligence community seeking increased access to unclassified imagery and information that can be readily shared with friends and allies.[120] Commercial imagery fits this need perfectly, contributing to the promulgation of truth and building of trust.

Government Space Science & Technology Budgets are Inadequate to Sustain U.S. Leadership - As a percentage of the federal budget, both civil space spending and military space spending are half the levels they were in the latter Cold War. Recent analysis suggests that the PRC is already at parity or outspending the U.S. defense budget. Overall DOD S&T is reported to be just 70% of Cold War levels [121] at just 1.9% of the DOD budget. But military space S&T is the lowest both in absolute dollars (about 1/10th the Army, Air Force and Navy) and as a percentage of total service budget (space S&T is just 0.9% vs Air Force 1.9%).[122] *Where the research budgets for other domains are measured in billions, space S&T is measured in mere millions, barely enough to tread water*, and at least four-fold too low to be competitive (see Table 1).[123]

DOD Science & Technology	Space Force	Air Force	Army	Navy (+Marines)
Total Budget	$15,300	$153,600	$178,000	$207,000
Basic Research (6.1)	-	$492	$463	$603
Applied Research (6.2)	$131	$1,400	$923	$953
Adv. Tech Development (6.3)	-	$779	$1,200	$760
Total S&T funding	$131	$2,700	$2,600	$2,300
S&T % of Total Budget	0.9%	1.7%	1.5%	1.1%

Table 1: FY21 DOD Science and Technology (S&T) funding by Service Branch (in $ millions).[124]

The Fundamentals have not changed - Six key areas, identified in the SSIB'20 report, *remain* central to democratic vs authoritarian competition for space security and U.S. and allied leadership, and America must compete in all:

- *Space policy and finance tools* to secure U.S. space leadership now and into the future by building a unity-of-effort within the government and incentivizing the space industrial base.
- *Space information services* including space communications/internet, PNT, and the full range of Earth observing functions which have commercial, civil and military applications.
- *Space transportation and logistics* to, in and from Cislunar space and beyond.

---

[120] B. Swan and B. Bender (2021). Spy chiefs look to declassify intel after rare plea from 4-star commanders. Politico.; Memorandum from USINDOPACOM signed by 9 Combatant Commanders, dated 15 Jan 2020.
[121] Cooley, W., Dougherty, G. (2021). Every Airman and Guardian a Technologist. ASPJ.
[122] Imperato, A., Garretson, P., & Harrison, R. (2021). U.S. Space Budget Report. AFPC.
[123] Garretson, P. (2021). Don't Skimp on Space S&T. Real Clear Defense.
[124] Imperato, A., Garretson, P., & Harrison, R. (2021). U.S. Space Budget Report. AFPC.

- ***Human presence*** in space for exploration, space tourism, space manufacturing and resource extraction.
- ***Power for space systems*** to enable the full range of emerging space applications.
- ***Space manufacturing and resource extraction*** for terrestrial and in space markets.

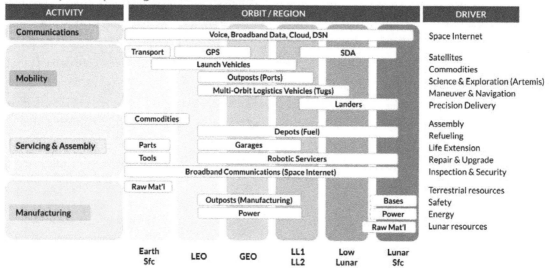

Figure 15: Drivers and key activities of a Cislunar economy enabled by an in-space logistics infrastructure.[125]

A Growing Consensus Views Space as Infrastructure - SSIB'21 participants conceptualized space as larger than just a government domain (exploration and national security) and more as infrastructure for commerce, communication, transportation and economic development. Existing space systems are increasingly being seen as critical national infrastructure in need of protection.[126] Future investments in transportation and industrial facilities to support an *in-space Cislunar econosphere* are likened to our nation's past investments in its canal system, Transcontinental Railway, national highway system, aviation infrastructure, and internet. Despite this growing consensus among space professionals and industry, the weak showing of space in the Endless Frontier Act[127] and Infrastructure Bill (INVEST in America Act[128]) is evidence that too few policymakers understand the potential payoff of space industrialization and development of space resources for in-space manufacture. Although the nation is contemplating a multi-trillion, once-in-a-generation infrastructure plan to 'win the future,' no part of it addresses the emerging in-space infrastructure or China's desire to surpass the U.S. in building it.

A Strong Bipartisan Consensus Supports U.S. Space Leadership - This is the first time we have seen such strong continuity across a change in administrations. The basic goals, programs, and governance structures have remained in place: the National Space Policy, National Space Council, NASA Artemis Program, Artemis Accords international initiative, and U.S. Space Force. NASA's

---

[125] Presentation by Steve Butow, DIU at SSIB'21
[126] Foust, J. (2021). House bill would designate space as critical infrastructure. SpaceNews.
[127] U.S. Congress (2021). S.1260 - United States Innovation and Competition Act of 2021. Congress.gov.
[128] U.S. Congress (2021). H.R.3684 - Infrastructure Investment and Jobs Act. Congress.gov.

Administrator, Senator Bill Nelson, recently addressed the National Space Symposium stating, "NASA's 21st century lunar exploration program will make new discoveries, advance technologies, and show us how to live and to work in another world. For that to be possible, the space program needs constancy. That's why NASA must be a nonpartisan agency, and why the Artemis program has bipartisan support. For the first time in more than 50 years, NASA will return humans to the Moon. We will go in a way that reflects the world today, with government, with industry, and with international partners in a global effort."[129]

> *"The National Space Strategy commits the Space Force and the Department of Defense to monitoring what's going on there [across Cislunar space] and if somebody is a bad actor in the international realm, to monitor, detect, and respond."*
>
> — DR. KELLY HAMMETT, AFRL, 22 June 2021

A Broader Vision of the Space Force Strategic Purpose is Emerging - Various think tanks have begun to champion a broader strategic purpose and expanded area of responsibility for the Space Force. CSIS,[130] CSPC,[131] AFPC,[132] Aerospace,[133] and the Atlantic Council have all provided recommendations for Cislunar, with former USSTRATCOM commander General Cartwright and former SECAF James championing a 'Cislunar Approach' and a 30-year strategy.[134] Earlier this year the Space Futures Workshop[135] explored how the expanding area of responsibility and increased activity are likely to drive expanded roles, missions and technology investments.

Figure 16: Renewed strategic competition presents a credible threat to the free flow of commerce in space (With permission of the artist: Jeff Koterba, http://jeffreykoterba.com)

---

[129] Jewett, R. (2021). Bill Nelson Emphasizes NASA's Climate Priorities at Space Symposium. Via Satellite.
[130] Kaplan, S. (2020). Eyes on the Prize: The Strategic Implications of Cislunar Space and the Moon. CSIS.
[131] CSPC (2021). Maintaining Momentum In National Security Space.
[132] Colucci, L. (2020). The Case for Space Development. AFPC.
[133] Vedda, J. (2018). Cislunar Development: What To Build—and Why. Aerospace.
[134] Starling, C. et al (2021). The Future of Security in Space: A Thirty-Year U.S. Strategy. Atlantic Council.
[135] U.S. Space Force. (2021). SpOC hosts 2021 USSF Space Futures Workshop. Spaceforce.mil. See Appendix B.

New Instruments are Reaching Non-Traditional Companies - The variety of new approaches to commercial tech discovery such as DIU, AFWERX, SpaceWERX and Space Prime are succeeding in broadening the number of companies' participation in the national security innovation base. These successes reinforce the necessity of organizations designed to accelerate the adoption of commercial innovation at a time when two thirds of our nation's research and development expenditures involve non-government sources.[136] The challenge remains that Requirements-based acquisition results in the vast amount of DOD procurement dollars going exclusively to traditional vendors focused on bespoke government solutions with little, if any, commercial nexus (see Figure 17). If the U.S. is to retain its technological leadership and compete globally, it must diversify its portfolio and increase the total percentage of commercial procurements as prescribed by Congress. This is keeping with the adage that the U.S. Government should "buy what it can, and only build what it must." In doing so, more buying power is achieved by negating design, sustainment and improvement costs.

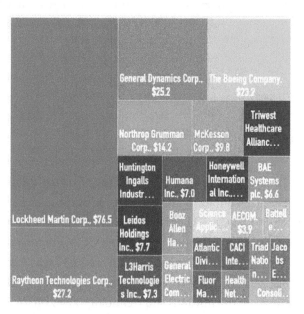

Figure 17: The top 25 defense contractors received 80% (or $235 billion) of DOD's procurement funding for bespoke government solutions in FY 2020 (Credit: DIU).[137]

"In the eyes of the world, first in space means first, period; second in space is second in everything."
— LYDON B JOHNSON, 1961[138]

ISSUES & CHALLENGES

While an impressive momentum exists in the Space Industrial Base, significant challenges constrain its potential:

STRATEGIC ISSUES

Space Policy is Not Connected with Domestic or Foreign Policy Priorities - Although space offers vast opportunities for economic growth and extension of U.S. soft power, the space industrial base has not been mobilized as a tool of domestic economic growth nor as a platform to advance American leadership on global problems and aspirations such as infrastructure diplomacy, science

---

[136] Sargent, J.F. (2020). U.S. Research and Development Funding and Performance: Fact Sheet. CRS.
[137] Data from USASpending.gov.
[138] McDougall, W. A. (1985) ...the Heavens and the Earth: A Political History of the Space Age. Basic Books.

diplomacy, and the most visionary climate solutions. For example, space is not featured in the articulation of national technology priorities, the infrastructure bill, is an afterthought in the Endless Frontier Act, is not a central pillar of the G7 infrastructure proposals, and is not part of the narrative of America's approach to climate change solutions (for example, in the Administration's Executive Order [139] on or the U.S. proposal for the United Nations' COP26 Conference[140]). For space to fully support the national agenda, it needs to be framed at the national level as a 21st century industry, modern infrastructure, national technological priority, and in particular *as a source of solutions for climate change*, green energy and sustainable development.

"It is the policy of my Administration to organize and deploy the full capacity of its agencies to combat the climate crisis to implement a Government-wide approach that reduces climate pollution in every sector of the economy; increases resilience to the impacts of climate change; protects public health; conserves our lands, waters, and biodiversity; delivers environmental justice; and spurs well-paying union jobs and economic growth, especially through innovation, commercialization, and deployment of clean energy technologies and infrastructure."

–PRESIDENT JOSEPH R. BIDEN JR., 27 Jan, 2021[141]

What's at Stake - is no less than whether the largest geographic zone of human activity is one of democratic freedom and fair trade, or an autocratic exclusion zone. Will citizens of the world conduct their activity under a presumption of freedom, human rights, rule of law and a rules-based-order, or will they be mere extensions of a totalitarian state using rule-by-law? Will the material resources of outer space—a million-fold what is on Earth accrue to totalitarian powers and their resource-nationalist tendencies, or will they accrue to U.S. and like-minded nations where a balance of power that favors freedom prevails? Will the industries and jobs and partnerships of the 21st century, including leadership on climate change and green technologies accrue to the ambitions of those benefiting from 'partocracy,' or will they be in the hands of free people?

"Don't Screw Up" - Proactively Pursue Strategic Territory

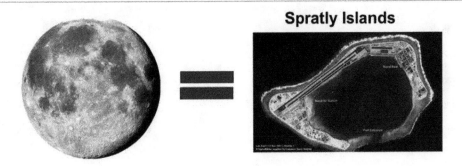

Figure 18: China will assert claims to Lunar territory and resources as it does with the Spratly Islands.[142]

---

[139] White House. Executive Order on Tackling the Climate Crisis at Home and Abroad. White HOuse.
[140] Reuters. COP26 must be "pivotal" to tackling climate change, says U.S. envoy. Reuters.
[141] McDougall, W. A. (1985) ...the Heavens and the Earth: A Political History of the Space Age. Basic Books.
[142] Presentation by Quilty Analytics at SSIB'21.

Cislunar Matters - Cislunar was a much greater part of SSIB'21 discussions that it had been in past workshops. According to Quilty Analytics, "More than 100 Lunar missions and 40,000 satellites are expected to be launched over the next decade" necessitating new infrastructure to identify, track, and communicate.[143]

It's Not a 100 Years Out! - Most policy-makers and the general public are unaware of the speed of innovation and competition. *Industry participants were clear that current visions and narratives are far too conservative and lack the necessary urgency.* Few are aware that in the last decade launch costs have dropped ten-fold, the number of active satellites in orbit have more than tripled.[144] Few are aware that in the next decade, *over 100 Lunar missions are planned*, in-space assembly and manufacture and power-beaming will be demonstrated, Lunar and asteroid mining will be demonstrated, sub-orbital rocket cargo will be demonstrated, over 40,000 new satellites launched, and at least 95 companies already have business plans involving Lunar or Cislunar.[145]

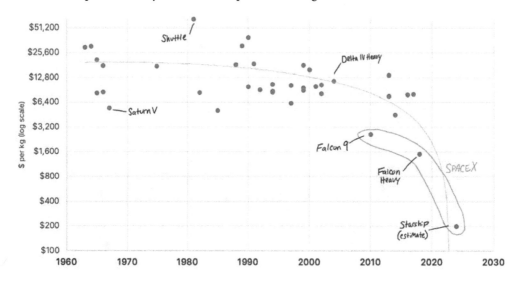

Figure 19: An explosion of startup activity with more than 90 companies pursuing cislunar activities.[142]

Figure 20: Medium and heavy space launch costs per kilogram to low Earth orbit[146]

---

[143] Presentation by Quilty Analytics to SSIB'21. See Appendix D.
[144] Mazareanu, E. (2021). Number of active satellites from 1957 to 2020. Statista.
[145] Presentation by Quilty Analytics to SSIB'21; Bryce S&T (Aug 2021). Projected Exploration Missions (2020-2030). Retrieved from https://www.bryce.com
[146] Data sourced from CSIS Aerospace Security. Graphic credit: FutureBlind.com.

They are not accounting for a "Starship Singularity"[147] where fully reusable heavy lift rockets will enable global transport in under 30 minutes, daily launches to orbit of 100 metric tons for $2-5 million at <$100/kg.[148] These same vehicles are being designed to refuel on orbit and to be able to land on all rocky planets and moons in the solar system.

"Zhèlǐ bù jiǎng yīngyǔ (这里不讲英语)"
[English not spoken here]

Flag of Choice in Jeopardy - Near term action or inaction may decide if English, the international language of aviation, will become the international language of astronautics and space traffic management, if GPS or a competitor will be the standard for precision navigation and timing in Cislunar space, if open internet and U.S. broadband are the standard for Lunar commerce or our allies and partners are forced to go with the infrastructure put down by others.

Figure 21: China's Zhurong rover next to its lander on Mars (Image credit: CNSA).[149]

Accelerating Soft Power Competition - The increasing capability of America's adversaries are allowing them to challenge U.S. leadership with alternate institutions for space exploration,[150] infrastructure development and international standards.[151]

---

[147] *"Starship Singularity"* phrase coined by Michael Laine, CEO of Liftport to describe the discontinuity represented by a system which is hugely capable (100MT to orbit), fully reusable and cheap ($2-5 million per launch, $10-20/kg), on-orbit refuelable, able to land on other moons and planets, and able to launch at a high cadence (weekly or daily) beyond which current assumptions and expectations cease to hold and predictions become unreliable.
[148] Zafar, R. (2020). Elon Musk Reiterates Insanely Low Starship Launch Costs Of $10/kg. WCCFTECH.
[149] Pultarova, T. (2021). China's Mars rover Zhurong just snapped an epic self-portrait on the Red Planet. Space.com.
[150] Leonard, D. (2018). China's Space Station Will Be Open to Science from All UN Nations. Space.com; Jones, A. (2021). China, Russia reveal roadmap for international moon base. SpaceNews.
[151] Hui, J. (2019). Programme and Development of the "Belt and Road" Space Information Corridor. CNSA.

Call the Play / Motivate the Team - America's space industrial base is the equivalent of a super power. It's talent and energies are vast. But it is like a disorganized team waiting on a team captain to call the play, coordinate and motivate the team to compete. Industry knows exactly what it wants to achieve but the USG is failing to recognize the full potential of what America's space industrial base can contribute to our national security and defense. The lack of national purpose, clarity, and strategic direction is having a deleterious effect, creating risk, uncertainty and decoherence among commercial actors. America's team is waiting to contribute to the national agenda, but they are not being acknowledged, tasked or motivated.

Guidelines for a North Star Vision - To take advantage of the significant opportunities presented by growth in the space industry, a White House level vision is needed whose aims and objectives are broadly supported across partisan lines and can survive multiple administrations. It needs to be well connected to enduring domestic and foreign policy goals. It must incorporate the central elements of the big audacious goals current in industry, partners and contenders: space economic development, a Cislunar economy, Lunar and asteroid mining, moving industry off Earth, space solar power, planetary defense, and space settlement, enabling humanity to become the first multi-planetary species. It must assume the nonlinearity of compound growth. It must set targets and timelines at least through 2050 (as that is the scope of the public plans and visions of our competitors). It must guard our supply chain. It must set production targets to energize industry. It must be whole-of-nation, include taskings for all relevant federal departments, engage the full diversity of America through empowering both individual states, and universities. It must provide a shaping and wealth-generating platform for America's allies and partners to improve their own conditions and link together like-minded nations. It must be connected to 21st century infrastructure, industries, jobs, and climate goals. It must outpace China and other autocracies. These guidelines are explored in depth in the 'North Star Vision' section.

TACTICAL ISSUES

The Lack of Meaningful ISR Contracts is Harming U.S. Companies - A broad spectrum of space-based Earth observation satellite companies have emerged capable of providing Incident Awareness and Assessment (IAA), peacetime pattern of life and indications and warnings, and tactical Intelligence, Surveillance and Reconnaissance. However, token investments are not sufficient, and the slow pace of commercial acquisition by the NGA and NRO is creating incentives for U.S. companies to withdraw from the national security track.

Overclassification is Harming Innovation - Most innovation comes from small businesses, including in the space sector. New companies often do not have the clearances or even personnel to handle the large tax levied by security requirements. Overclassification 'architects out' the potential contributions and solutions of innovative start-ups, making government contracts only available to established incumbents who have sufficient slack to pay to handle security requirements.

Licensing Bottleneck - Small companies face severe penalties when they lose contracted launch opportunities due to licensing delays. For example, numerous commercial companies were scheduled to fly on the SpaceX Transporter 1 flight this year. Despite applying months in advance, many commercial payloads had not cleared NTIA as part of the FCC licensing process. Those commercial payloads with direct DOD contracts were able to be resolved through DOD engagement, but others were not so fortunate and had their smallsats pulled from the flight, severely impacting start-ups for whom such delays and expenses 'make or break' their business cases.[152]

Transitioning the 'Valley of Death' - Although significant progress has been made in finding and maturing commercial solutions through AFRL, DIU and others, the limiting factor is the low level of subscription by acquisition services or programs of record which are often tied to 'requirements'-based acquisition. The DOD needs a central acquisition marketplace for commercial products and services drawn from the commercial sector.

	Pre-2010	New Entrants Post-2010		
	Legacy Players	Startups	Equity Raised ($M)	
Small Responsive Launch	16	74	$2,718	(1)
Launch Reusability	8	24	$1,570	(1)(2)
SAR Satellites	5	7	$629	
Flat Panel Antennas	1	19	$612	
Spacecraft Propulsion	15	37	$198	
Optical Intersatellite Links	4	25	$111	

(1) Relativity Space counted in both sections. Source: Company reports and Quilty Analytics.
(2) Excludes SpaceX, which has individually raised $6.4 billion.

Table 2: Commercial startup and funding activity of select space industry technologies.

Capital Misallocation - Although there is significant investment in commercial space, there is an over-concentration in launch and too little capital flowing into other verticals ready for investment. Government has effectively messaged its need for launch, but additional clarity and shift toward commercial purchase of goods and services is needed in order to spur investments across the industry.

Accelerating Hard Power Competition - America's adversaries continued to mature their counter-space threats and military space capabilities.[153] New intelligence estimates have highlighted these threats.[154]

Predatory Foreign Practices - Foreign Ownership Control or Influence (FOCI)[155] remains a significant concern for the space industry. U.S. companies continue to be targets of foreign predatory

---

[152] Foust, J. (2021). SpaceX launches record-setting cluster of smallsats. SpaceNews; and email correspondence with S. Butow, Dec 2020.
[153] Weeden, B. & Samson, V. (2021). Global Counterspace Capabilities. Secure World Foundation; Harrison, T. et al. (2021). Space Threat Assessment 2021. CSIS.
[154] ODNI. (2021). Annual Threat Assessment of the Intelligence Community. DNI.Gov.; OSD. (2020). Military and Security Developments Involving the People's Republic of China 2020.Media.defense. NASIC. (2018). Competing in Space. Media.defense.
[155] DSCA. (n.d.). Foreign Ownership, Control or Influence (FOCI). DSCA.mil

practices through malign investment and intellectual property theft.[156] Industry is under strain and at risk of being supplanted by the PRC for supply because of how PRC maneuvers in market places.[157]

No Industrial Targets for Artemis - The lack of clear production targets[158] for in-space industry for commercial partners associated with the Artemis Program limits its impact as a tool for space resource and economic development. A strategic call for the creation of in-space logistics infrastructure, both physical and digital, would otherwise benefit civil, commercial and military/national security space.

Anemic Allied International Partnering - Although there is broad intent to increase efforts at international partnership with allies and regional partners - including via commercial partnerships - actual efforts are anemic. Limited past successes, notably the International Space Station, were government led. Executing international commercial space partnerships is an actionable way to elevate the importance of working with allies. The first International Pitch Day and NATO Space Pitch Days are initial steps. Despite the significant potential of foreign commercial partners to advance allied space capabilities and to help cement geostrategic partnerships, legacy practices and over-classification severely hampers what is possible.

Figure 22: The International Space Station has been a pillar of int'l cooperation for more than 20 years (Credit: NASA)

Each COCOM should seek at least three commercial space partnerships per year.

---

[156] Carson, B. (2021). The FBI's warning to Silicon Valley: China and Russia are trying to turn your employees into spies. Protocol.

[157] Brown, M. & Singh, P. (2018). China's Technology Transfer Strategy: How Chinese Investments in Emerging Technology Enable A Strategic Competitor to Access the Crown Jewels of U.S. Innovation. DIU.

[158] Garretson, P. (2019). Why the next Space Policy Directive needs to be to the Secretary of Energy. The Space Review.

> *"The United States is the first country to have private companies taking private passengers to space. This is a moment of American exceptionalism. That's how we see it...It will be the ingenuity of all of our commercial partners to help us continue advancing to the next stage of our nation's space exploration."*
> — JEN PSAKI, White House Press Secretary, 2021[159]

KEY ACTIONS & RECOMMENDATIONS

From the above observations and additional inputs from the working groups, participants advocated for or expressed interest in 18 overarching recommendations for action:

ATTENDEE RECOMMENDATIONS FOR THE WHITE HOUSE & SPACE COUNCIL

1. **Establish "Space Development and Settlement" as our National "North Star" Space Vision** - The United States still requires a whole-of-nation vision and strategy for the economic and industrial development of space, to unite all elements of national power, and to attract like-minded allies and partners to a common wealth-creation framework. Central to this is the establishment of clear production goals for in-space and Lunar industrial facilities. Such a vision and strategy must be relevant to the global agenda, consonant with the spacefaring ambitions of the electorate, and have sufficient bipartisan support to endure multiple administrations. The full mobilization of America's diverse talent set requires big audacious goals which cause the nation to stretch, aid public diplomacy, and create the perceptions of the U.S. as a vibrant attractive partner. Space offers solutions for tackling climate change - not only in monitoring and modeling but in scalable energy solutions to create a green space power grid and source the materials for an electric economy. Re-issue a White House level vision and follow it up with an executive order specifying whole-of-nation roles, goals and deliverables on clear timelines.

> *"1st in space, first in everything, 2nd in space, second in everything."*
> — *WORKSHOP PARTICIPANT, 2021*

2. **Build Back <u>Beyond</u>: Incorporate the Moon into the Earth's Economic Sphere by Catalyzing the Space Superhighway** - Building back better means an expanded economic canvas for America in the one theater that offers a million times Earth's resources and a billion-times its energy. Government can catalyze the development of 21st century infrastructure, industries, and jobs by creating markets, providing incentives, protecting investments, expanding and enabling infrastructure. Transportation, information, and communications infrastructures are fundamental antecedents to much broader economic activity. Attendees urged a government role in seeding an infrastructure. It is essential for the

---

[159] Press Briefing by Press Secretary Jen Psaki, July 20, 2021

United States government to focus on catalyzing both the physical and digital logistics infrastructure which will engage and enable profitable space commerce.

3. **Sustain Funding for the Hybrid Space Architecture[160] as a Foundation for the Future Space Internet** - Congress provided critical funding in FY21 to seed a hybrid information network architecture in space capable of connecting disparate civil, military, commercial and allied systems in a secure environment that leverages the cloud, layered security, and low latency communications. The internet sector created 6 million jobs and contributed $2.1 trillion to the U.S. economy in 2018.[161] The U.S. must accelerate and sustain this activity to thwart China's ambitious goal to dominate the Space Internet.[162]

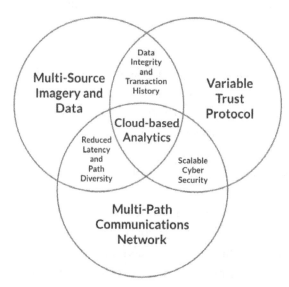

Figure 23: Proposed framework for a Hybrid Space Architecture connecting government, commercial and allied space systems in a secure, cloud-enabled internet-of-things environment.[163]

4. **Expand "Artemis Accords" Beyond NASA** - Establish international norms of behavior with Allies and partners to include operational space traffic management data exchanges, space domain awareness, and science and technology collaboration to promulgate democratic principles in space consistent with the rule of law, human rights, and liberties that underpin the norms, standards, and rules of acceptable behavior and action. This can be done deliberately by viewing U.S. Cislunar missions -- whether executed by NASA, DOD or commercial -- as opportunities for setting foundational precedents and standards.

5. **Increase Space Science & Technology Funding to Parity with Other Domains** - Neither our nation's spacefaring ambitions nor its economic or national security imperatives can be accomplished without technology leadership. All access to and accomplishments in space are enabled or mediated by technology. The competence of our adversaries and their pace of

---

[160] See Appendix F.
[161] Hooton, C. (2019). Measuring The U.S. Internet Sector: 2019. Internet Assn.
[162] Harding McGill, M. (2021). China's on a mission to dominate space internet. Axios.
[163] Presentation by Steve Butow, DIU at SSIB'21

innovation mean that we must restore overall space funding and military space S&T funding to past levels of strategic competition. Participants thought that given the centrality of space to strategic competition and integrated deterrence, space S&T would be funded at least to parity with the domains of land, sea and air. The speed of commercial innovation and continued U.S. commercial leadership is constrained at least, and more likely imperiled by the small size of the Space Force S&T budget.

6. **Reform Policy to Address 21st Century Conditions** - Most legacy policies were designed for a world with very different pressures and concerns. Technology and information ought to be a strength of the United States as a strategic partner. Yet legacy policies often result in the U.S. Government, it's partners and allies being slow or late to receive the latest and best technology from the very dynamic and vibrant U.S. industrial base. To counter this, the group recommended a specific example that included, relaxing commercial remote sensing regulatory restrictions to permit U.S. companies to compete and lead the world in commercially available capabilities.

7. **Declare Space a Special Economic Zone and Deploy the Full Range of Tools** - It is time to declare space a special economic zone and to deploy the full range of financial tools and incentives to stimulate growth and industrial expansion.

8. **Recognize Space-critical Infrastructure / Make Space a Part of Infrastructure Plans** - The fact that space systems constitute infrastructure critical to our way of life and prosperity is now broadly accepted, yet it is not formally recognized. Recognizing space as critical infrastructure opens new tools to sustain and grow the sector.

9. **Make Space a Central Part of Climate Action Plans** - The space industrial base is capable of immense and diverse contributions to tackling climate change. But it must be deliberately mobilized by White House policy. If appropriately directed, space holds not only better observations for models, but actual solutions at scale, the pursuit of which can allow the United States to own the international narrative of bold and audacious ideas in service of the global agenda.

10. **Include Space in Supply Chain Planning** - A failure to look ahead has put the U.S. behind on global infrastructure, 5G, rare-earth and medical supply chains. The U.S. space sector -- because it is the global leader -- is the subject of intense predation. As such, the Administration should prioritize space in its supply chain planning.

ATTENDEE RECOMMENDATIONS FOR THE DOD

"We're really looking to make a shift to a more defendable architecture. I think to do that, you will see where the hybrid architecture will be required."

— GENERAL JOHN RAYMOND, CSO, 14 December 2020[164]

11. **Integrate JADC2 with the Hybrid Space Architecture** - It is critical that the DOD leverages the capabilities of the Hybrid Space Architecture to fully enable Joint All Domain Command and Control (JADC2). Industry is ready now to contribute and accelerate advancement of the Joint All Domain Command and Control (JADC2) and other elements of the hybrid architecture with existing on-orbit capability. It is time to harness commercial capabilities via a hybrid space architecture to enhance the resilience of DOD space capabilities.

12. **Enable the Space Superhighway by Including Commercial Solutions for In-space Logistics Infrastructure** - As USSF articulates its architecture for in-space mobility and logistics, it is critical to include commercial solutions from the start. This effort should be done in close collaboration with NASA. In terrestrial modes of logistics, the ability to make use of civilian ports, civilian fuel, and civilian interfaces is a significant force multiplier -- the same is true in space. A logistics architecture that leverages commercial solutions will be more resilient, scalable, provide greater volume and reach, and see faster innovation because of private capital investment and its broader and more scalable economic impact.

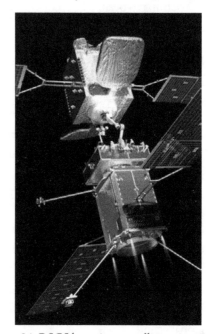

Figure 24: RSGS boosting a satellite to its mission designed orbit. (Credit: Northrop Grumman)

Enable in-space logistics infrastructure be accomplished by including commercial solutions in the DOD architecture, by forecasting an architecture which anticipates a mature Cislunar economy and associated new infrastructure, space lines of communication, and the availability of On-Orbit Servicing, Assembly and Manufacturing (OSAM) and by anticipating the availability of in-space resources such as propellant, and structural and functional materials.

---

[164] Aviation Week (2020). Podcast: Interview With U.S. Space Force Chief. Aviation Week.

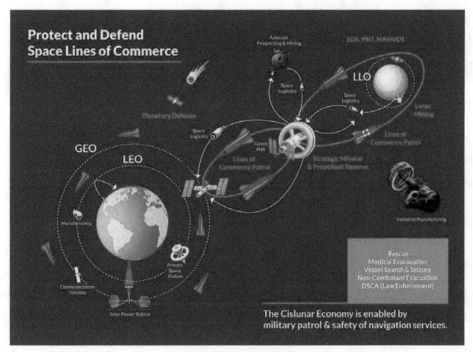

Figure 25: Protect and defend space lines of commerce (Credit: DIU).[165]

13. **Mandate a Percentage of Commercial Services Buys Starting in 2022** - Today commercial services procurements represent single digit percentages of the overall DOD acquisition budget.[166] The current policies, budgeting structure, and lack of procurement innovation incentives (plus perceived risk) contribute to this low level. Participants judged that a commitment to more agile and rapid innovation requires policy and incentives that drive toward a goal of 20% non-traditional commercial service acquisitions. This would greatly ameliorate the various technology valleys of death and enable the DOD to benefit from the much faster innovation cycles in industry.

"The Space Systems Command already has an office that procures commercial SATCOM. We are expanding the role of that office to look not only at commercial SATCOM but look at commercial services more broadly." – GENERAL JOHN RAYMOND, CSO, 24 August 2021[167]

---

14. **Expand Use and Management of Space Commercial Services within the Space Force** - The unprecedented levels of capital investment in space innovation and small business creation will not be sustained if investors cannot see revenue. The bar of 'significance' recognized by the venture capital community are contracts valued at $25 million. Currently DIU leverages $38 of private investment for every dollar of taxpayer funding, or $9.1 billion in private capital versus just $242 million in DOD funding through DIU. This group thought that a policy floor of 20% commercial / non-traditional purchases would most advance U.S. goals of agility, speed of

---

[165] Military defense of commerce roles and missions based on SSIB'21 presentation by Peter Garretson; lines of commerce and Cislunar econosphere components first depicted by ULA CisLunar 1000 at SpaceFlightNow)
[166] See USASpending.gov.
[167] Erwin, S. (2021). Raymond's progress report on Space Force: 'All the pieces are coming together'. SpaceNews.

innovation and acquisition reform. This could begin by requiring System Program Office (SPO) modernization programs to spend 20% of modernization funds on commercial service contracts (today commercial accounts for only 2% of acquisitions). These substantial programs dwarf the USSF S&T budget in scale and impact. DOD leadership must realize that a failure to award such contracts significantly increases mid- to long-term risk-to-mission, risk-to-force by contracting capital markets, turning innovative companies away from the national security track and potentially inadvertently incentivizing technology migration to peer competitor nations.

> *"The side that wins in the future is the side that has the greatest situational awareness and acts most quickly."*[168]
>
> -- GEN (RET) DAVID DEPTULA
> Mitchell Institute for Aerospace Studies

Participants recommended that procurement policy should create a presumption of purchase of data and services, and move towards a DLA-like model of specifying expected commodity buys via a space commodities exchange on the order of $1 billion. Participants noted that whether history records 2021-2022 as a bubble or a sustained slope of investment is not preordained, it depends on the actions of DOD leaders to send the appropriate signal for the future they want.

15. **Bolder Acquisition Reform Means a More Level Playing Field for All Business, Particularly Small Business** - Most innovation, economic growth and jobs come from small business, however structural barriers "architect out" many would-be commercial providers. Special requirements and modifications, security requirements, fiscal risk profiles, lengthy timelines to award, significant demand for meetings, paperwork-chase proposals severely disadvantage small businesses. Streamlining/simplifying the requirements process (single page Joint Requirements Oversight Council (JROC) level requirements), frequent and continued robust competition throughout product life cycles, embracing iterative DevSecOps approaches to software and hardware, disaggregation, and proliferation are keys to maintaining U.S. technology leadership and deterring peer adversaries from further weaponizing space.

16. **Enable Rapid Innovation by Shifting Resources from SBIRs to OTAs** - While Small Business Innovative Research contracts (SBIRs) can help in the very early stages of a small business, delivering real capability requires a shift to mid & late term Other Transaction Authorities (OTAs) led by DIU and SpaceWERX, which are more effective instruments than Phase II or Phase III SBIRs for nurturing the national security innovation base. But the rigidity of the SBIR funding allocation prevents the most effective maturation of small business commercial capabilities. SBIR can be thought of as a 'tax' on innovation resources, and 'tax relief' is in order. Participants recommended DOD sponsor legislative relief in FY2022 NDA to allow SBIR funds to be used for commercial solutions opening or equivalent OT Agreement.

---

[168] Wilson, J. (2017). What is global persistent surveillance? Military & Aerospace Electronics.

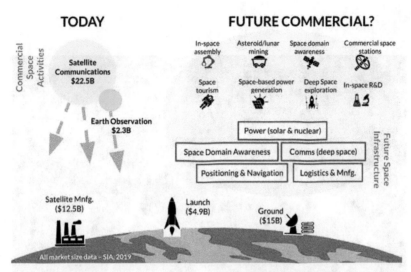

Figure 26: The growing space economy reflecting commercial space activities today and in the near future.[169]

ATTENDEE RECOMMENDATIONS FOR VENTURE CAPITAL AND INVESTORS

17. **Balanced Growth Requires Investment Beyond LEO** - The innovations in venture capital and SPACs have put the Low Earth Orbit (LEO) economy on solid ground with a tremendous diversity of space access options and space information services. However, we do not yet have a MEO, HEO, GEO, Cislunar or Lunar economy that is new space-oriented. This is despite the fact that there is clear interest from the government in expanding activity into these areas, a willingness to extend security into these areas, and no lack of capable start-ups with mature-enough technology seeking funding. To maintain growth, we need to diversify investment beyond LEO, expanding investments both in scale and to higher orbits. We want scale with the U.S. and its allies leading the way.

18. **Expand Investments in Enabling Technologies** - Key enabling technologies in the supply chain for the expanding space economy are missing or presently over-indexed off-shore. In particular, the supporting components such as microelectronics and scalable high-performance photovoltaics need to be developed and on-shored.

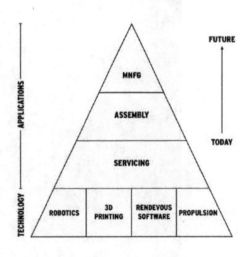

Figure 27: Elements of an in-space logistics and manufacturing market (Credit: Quilty Analytics).[170]

---

[169] Presentation by Quilty Analytics to SSIB'21.
[170] Presentation by Quilty Analytics to SSIB'21.

Figure 28: Illustration showing a close-up of the Habitation and Logistics Outpost (HALO), one of the elements of the Artemis Gateway launching no earlier than May 2024 (Credit: NASA).

A NATIONAL NORTH STAR VISION FOR SPACE

"In speaking of the rewards of space activity, we cannot ignore two others; these are the political and the spiritual. The United States with her leadership in advancing man's welfare— a leadership triggered by a superb science and technology— has become the standard that must be surpassed by any other nation if that nation is also to claim the distinction of leadership." – LLOYD BERKNER, 1960[171]

BACKGROUND

Unfinished Business - The top recommendation from last year's report remains a central concern of this year's event: "U.S. promulgate a "North Star" top-level vision and strategy for space industrial development and establish a Presidential Task Force to execute it (OPRs: POTUS, VP, NSpC, NSC, NEC, OTMP)."

Creating a National "North Star" Vision for Sustained Space Leadership - The United States still requires a whole-of-nation vision and strategy for economic and industrial space development to unite all elements of national power and to attract like-minded allies and partners to a common wealth-creation framework. Central to this is the establishment of a clear vision of a Cislunar economy by bringing the Moon into Earth's economic sphere with clear production goals for in-space and Lunar industrial facilities. Such a vision and strategy must be relevant to the global agenda, consonant with the spacefaring ambitions of

Figure 29: A North Star vision for civil, commercial and military space will unify all instruments of U.S. national security (Credit: NASA/Don Pettit)

the electorate, and have sufficient bipartisan support to endure multiple administrations. This is America's better answer to China's Belt and Road Initiative.

"The U.S. must develop and execute a grand strategy for space recognizing space's importance and enhancing our advantages. This strategy must encompass the near-term future, with space oriented as a source for augmenting terrestrial power, and the long-term future, encompassing space across the Cislunar expanse and beyond as a domain in itself for human action."
– USSF SPACE FUTURES WORKSHOP REPORT, 2021

---

[171] Wingo, D. (2015). The early space age: the path not taken but now? (Part 1). Retrieved from wordpress.com

Why This Is Important - Participants judged that without an integrated, comprehensive national space vision and strategy, U.S. space leadership and competitive advantage are at risk. Without a coordinated, consolidated national vision and policy there is insufficient demand signal to U.S. research and development agencies and the private sector. There is no national vision with articulated goals to 2050. There is also no national level vision that provides clear, time-specified goals which tie space power milestones to a broader long-term competitive industrial strategy and vision. Vision and narrative give meaning to our efforts and help allies interpret our initiatives. Vision - big ideas - and shaping platforms constitute a key part of soft power and public diplomacy. Vision enables mobilization of both national and allied energy and unity of effort. Space is seen as a surrogate for international leadership, societal attractiveness and vibrancy and national strength. Therefore, a clear national space vision is critical to enable whole-of-nation alignment and attract new partners.

Figure 30: A Dream Chaser launching a variety of missions to orbit (Credit: Sierra Nevada Corp)

Guidelines for a National North Star Vision: To take advantage of the significant opportunities presented by growth in the space industry, a White House-level vision is needed whose aims and objectives are broadly supported across partisan lines and can survive multiple administrations. It needs to be well-connected to enduring domestic and foreign policy goals. It must incorporate the central elements of the big audacious goals current in industry, partners and contenders: *space economic development, a Cislunar economy, Lunar and asteroid mining, the industrialization of space and moving industry off Earth, space solar power, planetary defense, becoming multi-planetary, and space settlement.* It must assume the nonlinearity of compound growth. It must set targets and timelines at least through 2050 (as that is the scope of the public plans and visions of our competitors). It must guard our supply chain. It must set production targets to energize industry. It must be whole-of-nation, and include taskings for all relevant federal departments. It must engage the full diversity of America through empowering both individual states and universities. It must provide a shaping and wealth-generating platform for America's allies and partners to improve their own conditions and link together like-minded nations. It must be connected to 21st-century infrastructure, industries, jobs, and climate goals.

Call the Play / Motivate the Team: America's space industrial base is the equivalent of a super power. It's talent and energies are vast. But it is like a disorganized team waiting in on a team captain to call the play, coordinate and motivate the team to compete. The lack of national purpose, clarity, and strategic direction is having a deleterious effect, creating risk, uncertainty and decoherence among commercial actors. The team knows they can contribute to the national agenda, but they are not being acknowledged, tasked or motivated.

Figure 31: Mars Base Camp is Lockheed Martin's vision for sending humans to Mars in about a decade. (Credit: Lockheed Martin)

CURRENT STATE

"The U.S. cannot ignore the potential of space as a major shaper of our present and future national power and the power of our rivals and adversaries. History shows nations who ignore new or expanding domains of human endeavor suffer for it. Other competitor and allied nations recognizing this potential are moving aggressively to position themselves in this future space world. The U.S. has advantages (historical, economic, political, and intellectual) we must exploit to meet these challenges."
— USSF SPACE FUTURES WORKSHOP REPORT, 2021

On July 23, 2020, the National Space Council released *A New Vision for Deep Space Exploration and Development*[172] which made some progress toward a whole-of-nation national 'North Star' vision, incorporating key elements highlighted in SSIB'21. At least one think tank has attempted to provide such a vision. Congress, in the most recent National Defense Authorization Act (NDAA), tasked the President and National Space Council to submit an assessment and strategy to compete with other national space programs and maintain leadership in the emerging commercial space economy.[173]

A Strong Bipartisan Consensus Supports U.S. Space Leadership - This is the first time we have seen such strong continuity across a change in administrations. The basic goals, programs, and governance structures have remained in place: the National Space Policy, National Space Council, NASA Artemis Program, Artemis Accords international initiative, and the U.S. Space Force.

A Broader Vision of the Space Force Strategic Purpose is Emerging - The Chief of Space Operation's General Raymond's planning guidance articulated that *"civil and commercial developments*

---

[172] U.S. National Space Council. (2020). A New Era for Deep Space Exploration and Development. Retrieved from Aerospace.org
[173] U.S. Congress (2020). Sect. 1614.Report and strategy on space competition with China, Public Law 116-283, National Defense Authorization Act for Fiscal Year 2021. Govtrack.

that pave the way for exploration and commercialization beyond near-Earth orbit will...require an order of magnitude expansion of our ability to sense, communicate and act to protect and defend American interests in cislunar space and beyond."[174] In a Senate hearing, General Dickinson, Commander, USSPACECOM, stated, "*we are opening our aperture to keep pace with our nation's expansion into the Cislunar region, to the Moon, Mars and beyond,*"[175] and later stated that Cislunar space is as strategically important as the high seas and likened the strategic importance of Cislunar Lagrange points to several very small islands of the Pacific.[176] His deputy commander, Lt Gen Shaw, outlined what it means for Space to be an Area of Responsibility (AOR).[177]

Figure 32: U.S. Space Force recruiting ad (Source: USSF).

Various think tanks have begun to champion a broader strategic purpose and expanded area of responsibility for the Space Force. CSIS,[178] CSPC,[179] AFPC,[180] Aerospace,[181] and the Atlantic Council have all provided recommendations for Cislunar, with former USSTRATCOM commander General Cartwright and former SECAF James championing a 'Cislunar Approach' and a 30-year strategy.[182] Earlier this year the Space Futures Workshop[183] explored how the expanding area of responsibility and increased activity are likely to drive expanded roles, missions and technology investments, concluding that "*USSF is committed to its broader strategic purpose to support space as a growing element of U.S. national power*" and that "*the Space Force will 'be there' wherever U.S. commercial and strategic interests and activities expand.*" Moreover, it anticipated that "*by 2040, USSF missions may include: increased space information services; projection of offensive and defensive operations in space and from space to other domains; dynamic offensive/defensive operations and transport across the Cislunar domain to ensure*

"Asteroids are nature's way of asking: 'How's that space program coming along?'" [184]
-- *NEIL DEGRASSE TYSON*
Director, Hayden Planetarium

---

[174] U.S. Space Force (2020). Chief of Space Operations' Planning Guidance. Defense.gov.
[175] USSPACECOM. (2021). USSPACECOM commander discusses space domain awareness, operating environment of space at Senate hearing. Spacecom.mil.
[176] Hitchens, T. (2021). USSPACECOM Head Touts Space, High Seas Parallels. Breaking Defense.
[177] Shaw, J. (2021). Space as an AOR: Speech by Lt Gen Shaw to Space 2021 Space Symposium. YouTube.
[178] Kaplan, S. (2020). Eyes on the Prize: The Strategic Implications of Cislunar Space and the Moon. CSIS.
[179] CSPC (2021). Maintaining Momentum In National Security Space.
[180] Colucci, L. (2020). The Case for Space Development. AFPC.
[181] Vedda, J. (2018). Cislunar Development: What To Build—and Why. Aerospace.
[182] Starling, C. et al (2021). The Future of Security in Space: A Thirty-Year U.S. Strategy. Atlantic Council.
[183] U.S. Space Force. (2021). SpOC hosts 2021 USSF Space Futures Workshop. Spaceforce.mil.
[184] Reynolds, G. (2013). Asteroids a reminder of space program weakness: Column. USAToday.

freedom of civil, commercial, military operations; environmental monitoring, stewardship and debris clean-up; protection of critical space national infrastructure; enforcing space law and norms of behavior; Search and Rescue / Personnel Recovery (PR) / Non-Combatant Evacuation (NEO); and planetary defense." [185]

Congressional Invitation to Expand Scope and Scale - The Congressional tasking provides an opportunity to broaden the national space vision animating U.S. space activity and strategic direction. The NDAA tasking is clearly not only about exploration, but asks specifically for *"a comprehensive assessment between the United States and China"* not only of human exploration and spaceflight, current and future space launch capabilities, but also to assess *"the strategic interest in and capabilities for Cislunar space"* and *"the viability and potential environmental impacts of extraction of space-based precious minerals, on-site exploitation of space-based natural resources, and the use of space-based solar power."* The NDAA task echoes many of the concerns in this report regarding the extent of foreign investment in the commercial space sector; theft of United States intellectual property; efforts by China to seize control of critical elements of the United States space industrial supply chain and United States space industry companies, and China's efforts to reach cooperative agreements with other nations to advance space development. The requested strategy is comprehensive in scope, seeking to maintain leadership in the emerging commercial space economy, to leverage commercial space capabilities to ensure the national security of the United States and the security of interests of the United States in space, as well as to protect supply chains, and coordinate with international allies and partners in space.

NOTE: Thirty-Five minerals or mineral material groups have been identified by USGS as critical minerals essential to the economic and national security of the United States, and the supply chain of which is subject to disruption. These include:

Essential Minerals (Group)	Moon Si Scale	Moon to Earth Si Ratio	Notes
Aluminum	528,700	1.7	Metals makes up about 10 to 18% of lunar regolith
Titanium	12,800	1.4	Abundant in the mare basalts (observed dark spots)
Chromium	3,490	18.0	
Zirconium	79.3	0.4	
Uranium	0.165	0.2	
Platinum	-	-	
Germanium	-	-	Abundance distributed. 300x larger at Apollo 15 site

Table 3: Lunar abundance of critical minerals essential to U.S. economic and national security.[186]

---

[185] U.S. Space Force. (2021). SpOC hosts 2021 USSF Space Futures Workshop. Spaceforce.mil.
[186] Taylor, S. (1982). Planetary Science: A Lunar Perspective. Lunar and Planetary Institute.

KEY ISSUES & CHALLENGES

Space Policy Must Match and Reflect U.S. Societal Ambitions - Private investment in space led by space entrepreneurs such as Jeff Bezos, Elon Musk, Sir Richard Branson, and others are leading to breakthroughs in cost, efficiency, and creative solutions across the space field. The entrepreneurs are driven by their shared belief in the expansion of humanity into space and the eventual settlement of the solar system. Bezos dreams of "moving all heavy industry off Earth" so that Earth can be "zoned residential and light industry"

Figure 33: The Moon will likely be our first permanent offworld residence (Credit: SpaceX).

and "a world for his great-grandchildren's grandchildren where humanity moves out into the solar system."[187] Musk aspires to colonize Mars and make humanity a multi-planetary species.[188] They both speak of millions of people living in space, and of scales far more ambitious than articulated in national vision or policy. Some may question their idealism in favor of other rationales, but their impact on U.S.' and allies' space technology advances and U.S. commercial space advantage continues to be dramatic - even more so their impact on general public perception and enthusiasm for the future of space and for continued U.S. space leadership. A national vision that fails to incorporate these visions will seem anemic and tone-deaf to the American electorate.

Current U.S. Space Policy Vision Must Match or Exceed the Scale of Adversary Visions and Narratives - The PRC has articulated and committed itself publicly to ambitious goals for space industrialization, space infrastructure and economic development, and exploration through 2050. These goals provide clear objectives and dates, including an international Lunar research station, construct "ultra-large spacecraft spanning kilometers,"[189] a 300 metric ton prototype power satellite,[190] the capture and return to Earth of an asteroid,[191] the industrialization of the Moon to build[192] 10,000 metric ton solar power satellites[193] at scale, reusable rockets for space access and point-to-point transportation, nuclear powered space shuttles for asteroid mining and space settlement,[194] and a Moon-Earth Economic zone with an annual return of 10 trillion dollars.[195]

---

[187] Clifford, C. (2018). Jeff Bezos Dreams of a World with a Trillion People Living in Space. CNBC.
[188] Statt, N. (2016). Elon Musk Says the Only Reason He Wants to Make Money Is to Colonize Mars. The Verge.
[189] Williams, M. (2021). China wants to build a spaceship that's kilometers long. Phys.org
[190] Jones, A. (2021). China's super heavy rocket to construct space-based solar power station. SpaceNews.;Chen, S. (2011). Space agency looks to capture sun's power. South China Morning Post
[191] Liangyu. (2018). China Focus: Capture an asteroid, bring it back to Earth?. Xinhuanet.
[192] Whittington, M. (2016). China plans to build space base solar power stations. Blastingnews.
[193] Jones, A. (2021). China's super heavy rocket to construct space-based solar power station. SpaceNews.
[194] He, T. (2017). China sees 'breakthrough' in nuclear-powered space shuttles by 2040. Global Times.; Chen, S. (2017). China's nuclear spaceships will be 'mining asteroids and flying tourists' as it aims to overtake U.S. in space race. South China Morning Post.
[195] Siqi, C. (2019). China mulls $10 trillion Earth-moon economic zone. Global Times.

Accelerating Soft Power Competition and Global Influence - Bold vision, robustly executed, attracts partners, talent, and capital. China has expended significant political capital committing to big audacious goals to aid its public diplomacy, soft power, attractiveness as a partner, and international prestige. The PRC mobilized its national energies toward their fulfillment of their announced goals on time, and this year it succeeded in efforts announced over a decade ago: Lunar sample return,[196] achieve Mars orbit, land, and deploy a rover,[197] launch and crew it's competitor space station[198] and test suborbital transportation systems.[199]

Figure 34: Commemorative plates are awarded to representatives from the international cooperators for the Chang'e-5 Lunar mission at the National Astronomical Observatories of China held in Beijing, China in January 2021 (Credit: Xinhua/Jin Liwang).

As a result of both its aspiration and success, China is in a strong position to draw new partners to participate in its space efforts, including its space station, Lunar research station, and Belt and Road Initiative Space Information Corridor. The increasing capability of America's adversaries is allowing them to challenge U.S. leadership with alternate institutions for space exploration, infrastructure development and international standards. China's aspirations and track record have also succeeded in attracting a surge in investment in its commercial sector.[200]

The National Space Vision Must Include Climate Change - While the perspective offered by space of Earth is broadly acknowledged[201] to have started the global environmental movement, few policy makers think of space itself as a green technology. Few policy makers seem aware that the improved route planning made possible by GPS reduces global transportation emissions between 15 and 21%.[202] It is only through space that we even became conscious of

Figure 35: Today's Earth science satellite fleet is being augmented with commercial remote sensing capabilities (Credit: NASA).

---

[196] Clark, S. (2021). Chinese mission returned nearly 4 pounds of Lunar samples. Spaceflightnow.
[197] Webb, S. & Allen, R. (2021). On its first try, China's Zhurong rover hit a Mars milestone that took NASA decades. The Conversation.
[198] Amos, J. (2021). China space station: Shenzhou-12 delivers first crew to Tianhe module. BBC.
[199] Jones, A. (2021). China launches secretive suborbital vehicle for reusable space transportation system. SpaceNews.
[200] Patel, N. (2021) China's surging private space industry is out to challenge the US. MIT Technology Review; Waidelich, B. (2021) China's commercial space sector shoots for the stars. East Asia Forum.
[201] Frank, A. (2018). *Light of the Stars: Alien Worlds and the Fate of the Earth*. WW Norton & Company.
[202] Autry, G. (2019). Space Research Can Save The Planet—again. Foreign Policy.

climate and its changes — it was efforts to model other planetary surfaces and understand heliophysics which led to the first climate models.

The Current Vision Fails to Articulate a Wealth-Creation Framework for Like-minded Allies and Partners - Present-day strategic competition in space is not principally about prestige or ideology signaling but about building an enduring wealth-creation platform to enable sustained growth and attract enduring alliances and partnerships. A central challenge for the U.S. is defeating our adversaries' efforts to pry away our allies and partners through offers of joint participation in the development of global platforms and international infrastructure for wealth, including space development. To reverse the weakening of the post-WWII order, the U.S. should deepen our ties with allies and partners who share our vision of a free and open space domain through creating a meaningful alternative that moves beyond just space exploration and military cooperation to cooperation in space development providing a path toward prosperity from an expanded space economy, secured through the stabilizing presence of the USSF.

The Challenge of a Vision that Sustains Across Administrations and Party Lines - The challenge before America in strategic competition is to sustain a consistent program across administrations. Our strategic competitors are able to sustain consistent vision and progress. This is the first presidential-transition where major space policy and goals have sustained across a transition in party leadership. However, the existing vision and its elements are insufficient to sustain American leadership, and broader vision is required. To serve the nation, it must be socialized on both aisles of Congress so that its central elements endure.

What's at Stake - is no less than whether the largest geographic zone of human activity is one of democratic freedom and fair trade, or an autocratic exclusion zone. Will citizens of the world conduct their activity under a presumption of freedom, human rights, rule of law and a rules-based-order, or as mere extensions of a totalitarian state using 'rule-by-law'? Will the material resources of outer space—a million-fold what is on Earth accrue to totalitarian powers and their resource-nationalist tendencies, or will they accrue to U.S. and like-minded nations where a balance of power that favors freedom prevails? Will the industries and jobs and partnerships of the 21st century, including leadership on climate change and green technologies accrue to the ambitions of those benefiting from 'partocracy,' [203] or will they be in the hands of free people?

Figure 36: Strategic competition with China could result in U.S. ceding its leadership role in space (Credit:techstartups.com)

Both because of the vast resources of space and the fact that key strategic terrain offers economic and force multipliers, Cislunar space appears to be a place where the Matthew Effect will prevail -- where advantage begets further advantage as those who are successful are most likely given special

---

[203] Goswami, N. (2021). China under Xi: The Institutionalization of the Communist Party of China. Live Encounters.

opportunities that lead to further success, while those who aren't successful are most likely to be deprived of such opportunities.[204] Strategic advantage in space compounds over time analogous to compound interest in a bank. Therefore, initial conditions matter, and create a path dependence for all future participants. The nation that emerges as the leader will set the precedents that condition the system, determining the rules of the playing field. Action therefore is urgent if we desire to retain U.S. leadership and secure for our children and grandchildren a second American century.

Incorporating Climate Change into the U.S. National Vision - Space has more to offer than just satellites for better climate data and models. Space offers solutions in several broad categories:

1. **Superior Climate Situational Awareness** - Space offers the vantage point to see and understand the world as a climate system. New satellite constellations -- especially commercial constellations -- present new opportunities for better sensing, more frequent sensing, and new sensing, including sensing of the upper atmosphere through radio occultation. Models and artificial intelligence pioneered in the private sector can provide far better resolution on emissions, mitigations, and effects. Space allows us to observe not only changes in temperature but changes in atmospheric gas composition. Space allows us to see real-time changes in the solar constant, and how these changes affect climate systems on Earth, Mars, and Venus to improve our climate models. An architecture which emphasizes modularity will allow us to add or subtract sensors as our understanding improves.

2. **Compliance Monitoring** - Space-based sensors offer the ability to monitor compliance of both domestic and international actors with regard to emissions, preservation of carbon sinks, and offsets such as deployment of new energy sources and planting of carbon sinks.

3. **Materials to Support the Green Economy** - The Moon and asteroids contain vast quantities of the rare-Earth elements to build efficient electronics, electric vehicles and hydrogen fuel cells. An off-Earth mining and manufacturing capability would enable a U.S.-controlled supply of strategic minerals — all with lower environmental impact to the biosphere.

4. **Avoiding Industrial Impact** - Over time, significant industry could be moved outside the biosphere where it would have much less impact on climate. For example, the insatiable appetite for data and processing is driving significant energy usage (~2% of electricity[205]), increased emissions and stressful cooling requirements. Some companies are already exploring how significant data processing could eventually be moved off Earth where solar power is strong and constant, removing its contribution to climate emissions.

5. **Controllable and Reversible Emergency Climate Interventions** - As proposed by Democratic Candidate Andrew Yang, developing an off-Earth mining and manufacturing capability could enable an "Earthshade"[206] at the Sun-Earth Lagrange point 1 could controllably reduce the solar

---

[204] Briggs, S. (2013). The Matthew Effect: What Is It and How Can You Avoid It In Your Classroom?. InformED.
[205] Shehabi, A. et al. (2016). United States Data Center Energy Usage Report. Berkeley Lab.
[206] Remarks by SpaceFund founder Rick Tumlinson at SSIB'21 plenary

constant by as much as 1.8%. This could buy time for a transition to a green economy without the potentially irreversible effects of other proposed geo-engineering solutions.

6. **Vast Reservoirs of Green Energy** - Lastly and most importantly, space has "big ideas" for green energy generation that deserve to be part of the conversation.

   a. **Helium-3 Fusion** - One motivation of both China and India to explore the Moon is its vast potential energy supply of a clean fusion fuel, Helium-3. Both China and India must take care to find energy resources which can sustain populations over a billion. China and India are betting that humanity will crack both Lunar mining and fusion, providing a resource large enough to power their billion person economies for centuries.[207] As stated by the Chief Chinese Lunar designer Ouyang Ziyuan, "*The Moon could serve as a new and tremendous supplier of energy and resources for human beings. This is crucial to sustainable development of human beings on Earth...Whoever first conquers the Moon will benefit first.*"[208] The United States, which first discovered the Helium-3 resource, and explained its significance to the world,[209] should not cede the field. Some level of interest exists in academia and the private sector to explore Lunar Helium-3 mining,[210] and the private sector is also exploring fusion propulsion for fast and safe interplanetary space transportation.[211] Yet Helium-3 has yet to be considered in an "all of the above" energy strategy.

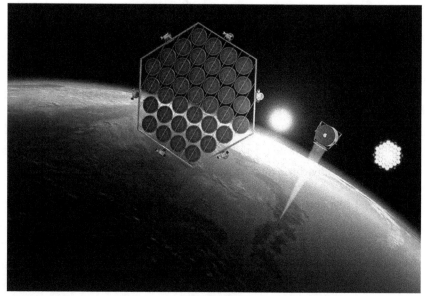

Figure 37: Space Solar Power (SSP) concept involving large solar collectors in GEO with the means of converting reflected light into microwave energy that is beamed to the Earth's surface by John MacNeill.[212]

---

[207] Goswami, N., & Garretson, P. A. (2020). *Scramble for the Skies: The Great Power Competition to Control the Resources of Outer Space*. Lexington Books.
[208] Whitehouse, D. (2002). China denies manned Moon mission plans. BBC.
[209] Schmitt, H. et al. (2011). Lunar Helium-3 Fusion Resource Distribution. Lunar Planetary Institute.
[210] Whittington, M. (2021). Solving the climate and energy crises: Mine the Moon's helium-3?. The Hill.
[211] Holland, A & Lintner, S. (2021). Fusion can take us farther faster. Madison.com.
[212] Illustration by John MacNeill with permission of the artist.

> *"It is time to reconsider SSP as a valuable tool in the nation's decarbonization strategy...the federal government should earmark approximately $1 billion for SSP research over the next five years with a special emphasis on advancing emerging technologies and in-space hardware demonstrations...We believe that a public-private SSP program...could result in a commercially viable SSP platform in geostationary orbit by the end of the decade."* – PROGRESSIVE POLICY INSTITUTE, 2021[213]

b. **Space-Based Solar Power** - The most ambitious and scalable idea in space, energy and climate is space solar power[214] or "*Astroelectricy.*"[215] It means constructing large orbital solar farms that collect the intense solar energy above the clouds and where there is no night, and transmit the energy wirelessly to the ground. If constructed using the materials of the Moon or asteroids, these power stations could scale to all global demand many times over,[216] with the International Academy of Astronautics assessing that "*annual employment on the order of 5,000,000 individuals might be realized eventually.*" This is an idea invented in America by Dr. Peter Glaser, and recommended for action by the Pentagon study group over a decade ago.[217] China currently leads the world in space solar research and development. Its big audacious goal to construct (circa 2030) "the most ambitious space project in history" -- a 300 metric ton multi-megawatt solar power satellite prototype designed to "*span at least one square kilometre, dwarfing the International Space Station and becoming the biggest man-made object in space.*"[218] The prototype is just a waystation in China's vision to industrialize the Moon to build solar power satellites[219] and establish a 'Moon-Earth economic zone' which will return $10 trillion annually in 2050.[220] Success would mean leadership in yet another of the key technologies and industries of the 21st century, the capture of up to five million new jobs and a chokehold on global energy. Because solar power satellites have the largest requirements for raw materials in space, leadership in solar power satellites likely also means leadership (and control) across multiple strategic sectors, including in-space Lunar and asteroid mining, in-space manufacture, space robotics and space logistics and transport. The mere audacity of such goals and their relevance to the global green and sustainable development agenda are sure to enhance its soft power and attract partners. Explored and then abandoned decades ago by NASA and DOE, today there is no national program, despite calls from both policy think tanks and media.[221] Space solar power was recently highlighted by the Aerospace Corporation's Space Agenda 2021 as a near-term investment decision before the nation[222] and

---

[213] PPI (2021). The Next Frontier Of Renewable Energy Is In Space, Argues New Report From Ppi's Innovation Frontier Project.
[214] Department of Energy. (n.d.). Space-Based Solar Power. Energy.gov.
[215] Snead, M. (2019). Rationale for a national "astroelectricity" program. The Space Review.
[216] Garretson, P. (2017). Solar Power in Space?. SSQ.
[217] National Security Space Office (2008). Space-Based Solar Power: As an Opportunity for Strategic Security.
[218] Jones, A. (2021). China's super heavy rocket to construct space-based solar power station. SpaceNews.; Chen, S. (2011). Space agency looks to capture sun's power. South China Morning Post.
[219] Xinhua. (2016). Exploiting earth-moon space: China's ambition after space station. China Daily.
[220] Siqi, C. (2019). China mulls $10 trillion Earth-moon economic zone. Global Times.
[221] AFRL. (2020). Space Power Beaming: Space Solar Power Incremental Demonstrations and Research Project (SSPIDR).
[222] Vedda, J. & Jones, K. (2020). Space-based Solar Power: A Near-term Investment Decision. Aerospace.

the Progressive Policy Institute has recommended the federal government allocate $1 billion over five years toward a prototype as part of America's decarbonization strategy.[223] It was also recently nominated by the largest space advocacy organization, the National Space Society, for a COTS-type Public Private Partnership.[224] The most ambitious proposal has been championed by the International Association of Machinists and Aerospace Workers (a.k.a. the Machinists' Union) to include a multi-billion Solar Power Satellite program as part of the infrastructure bill.[225] Recent interest in Space Solar Power has been expressed by friendly countries including India,[226] Japan,[227] Australia,[228] Canada,[229] the UK,[230] and UAE[231] as a potential way to honor their Paris climate emissions commitments, but the U.S. has no platform to coordinate such a large infrastructure effort. What exists today are just incremental technology demonstrator programs, the AFRL SSPIDR[232] program, the NRL PRAM-FX experiment,[233] a new $100 million private grant to CalTech,[234] and a few start-ups unable to make progress because of the lack of a SpacePrime or COTS-like effort to draw in investors. Incorporating space solar power into America's space and climate agenda could not only provide yet another arrow in the quiver to address climate change but provide novel ways to engage industry, the public, and international partners.

"United States the only major space faring nation whose national space agency does not have a serious plan to develop a SSP platform...Given SSP's benefits and the interest in the technology from most other space agencies, it's puzzling that policymakers in the United States have not prioritized SSP R&D."
— PROGRESSIVE POLICY INSTITUTE, 2021[235]

---

[223] PPI (2021). The Next Frontier Of Renewable Energy Is In Space, Argues New Report From Ppi's Innovation Frontier Project.
[224] National Space Society. (2020). A Public/Private COTS-Type Program to Develop Space Solar Power. NSS.
[225] Myers, E. (2021). *How President Biden, Congressional Democrats, NASA and the 2021 Infrastructure Bill Can Unite America and Achieve the Job Creation, Climate Change and Social Justice Goals of the Green New Deal*. International Association of Machinists and Aerospace Workers
[226] Rajagopalan, R. (2011). Space Based Solar Power: Time to Put it on the New U.S.-India S&T Endowment Fund. ORF.
[227] JAXA (2021). Research on the Space Solar Power Systems (SSPS). JAXA
[228] Zappone, C. (2019). Australia leans into space race for solar power with China. Sydney Morning Herald.
[229] Space Canada. (n.d.). About Space Canada.
[230] UK Government (2020). UK government commissions space solar power stations research. Gov.UK.
[231] Gopalaswami, R. (2015). UAE homes can get power from space. Khaleej Times.
[232] AFRL. (2020). Space Power Beaming: Space Solar Power Incremental Demonstrations and Research Project (SSPIDR).
[233] Leonard, D. (2021). Space-based solar power getting key test aboard U.S. military's mysterious X-37B space plane. Space.com.
[234] Potter, N. (2021). Solar Power from Space? Caltech's $100 Million Gambit Billionaire makes secret donation for electricity from orbit. IEEE Spectrum.
[235] PPI (2021). The Next Frontier Of Renewable Energy Is In Space, Argues New Report From Ppi's Innovation Frontier Project.

Make Space Development and Settlement the U.S. National Vision - The full mobilization of America's diverse talent set requires big audacious goals which cause the nation to stretch, aid public diplomacy, and create the perceptions of the U.S. as a vibrant attractive partner. Space offers solutions for tackling climate change -- not only in monitoring and modeling but in scalable energy solutions to create a green space power grid and source the materials for an electric economy. Participants suggested the administration re-issue a White House level vision and follow it up with an executive order specifying whole-of-nation roles, goals and deliverables on clear timelines. This development should contain two key elements:

- **Build Back <u>Beyond</u>: Incorporate the Moon into the Earth's Economic Sphere** - Building back better means an expanded economic canvas for America in the one theater that offers a million times Earth's resources and a billion-times its energy. The vision should include production targets for one or more public-private Lunar industrial facilities starting at the South Pole, and prioritizing national investments toward the green energy concepts discussed above.

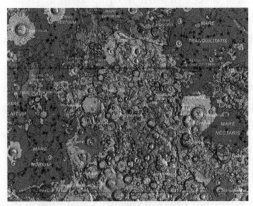

Figure 38: Unified Geologic Map of the Moon, 1:5M (2020) Source: USGS

- **Catalyze the Space Superhighway Infrastructure and Logistics Initiative** - Government can catalyze the development of 21st-century industries and jobs by creating markets and providing infrastructure. Transportation and communication infrastructure are fundamental antecedents to much broader economic activity. Attendees saw a government role in seeding an infrastructure. It is essential for USG to focus on catalyzing the logistics infrastructure which will engage and enable profitable space commerce.

- **Anchor Tenancy and Buying Commercial** -- Participants asserted the U.S. Government must help create the market and drive demand by purchasing commercial products and services, especially those which have the opportunity to scale to a broader customer base such as power, propellant, habitats, and strategic minerals.

KEY INFLECTION POINTS

> *"We are in a space race with China. They are aggressive. They are good."*
> — SENATOR BILL NELSON, NASA Administrator

- **U.S. fails to author a national vision consistent with its societal ambition and comparable to its strategic rivals** and thereby abandons leadership of the field to be eclipsed by more forward leaning governments.

- **U.S. succeeds in authoring a national vision consistent with its societal ambition and comparable to its strategic rivals** and is able to energize its commercial sector and signal to its allies its willingness to provide visionary leadership and action in the space domain.

- **U.S. fails to socialize its national vision to survive a partisan change in administration** and the brief gains of one administration prove ephemeral; the see-saw of partisan change retards progress, and the U.S. falls behind America's rivals.

- **U.S. succeeds in socializing its national vision to survive a partisan change in administration** and thereby is able to maintain focus, momentum and leadership to shape and structure the space domain.

- **U.S. industry makes the shift to a modular space design paradigm**, improving the speed of technological introduction, bolstering the supply chain, leveraging a wider class of launch vehicles, and improving efficiency and affordability, all of which enhance U.S. leadership in space.

Figure 39: The Augmentation System Port Interface (ASPIN) can dock to existing satellites and add new capabilities to a spacecraft that is already in orbit (Credit: Lockheed Martin).

> *"The U.S. must develop and execute a grand strategy for space recognizing space's importance and enhancing our advantages. This strategy must encompass near-term terrestrial-focused power and a long-term focus on Cislunar expansion and beyond as a domain in itself for human action."*
> — USSF SPACE FUTURES WORKSHOP REPORT, 2021

KEY ACTIONS & RECOMMENDATIONS

SHORT-TERM PAYOFF

Update White House's *A New Vision for Space Development and Exploration* with big audacious societal goals for development, settlement, use, and provide clear timelines and aspirational goals for identifiable and measurable outcomes and actions by U.S. agencies. (OPRs: VP, NSpC)

Socialize the National North Star Vision with Congress and task the heads of NASA, USSF, DOC, DOT to write OpEds to the electorate providing a mutually supporting narrative with an aim to sustain the broad societal goals across administrations. (OPRs: VP, NSpC)

Publish an Executive Order and Energize the National Space Council by providing specific tasking to all federal agencies. (OPRs: VP, NSpC, NSC, OSTP)

Stand up a National Space Enterprise Task Force to serve as an interagency coordinating body, synthesizing diverse perspectives and rapidly resolving space-related issues across participating departments and agencies. (OPRs: NSpC, NSC, NEC, OTMP)

MID-TERM PAYOFF

Lead International Efforts to Create International Institutions for the Space Economy analogous to the post-WWII building of the transatlantic community and global economy such as the Atlantic Charter, Organization for Economic Cooperation and Development (OECD), the International Civil Aviation Organization (ICAO), and Bretton-Woods institutions including the World Bank, International Monetary Fund (IMF). Using the vision, this becomes America's global wealth creation platform, analogous to the efforts such as the U.S.' Marshall plan, and China's Belt and Road Initiative. (OPRs: EOP, NSpC, DOS)

Construct a Space Defense Alliance to Secure the Vision analogous to the North Atlantic Treaty Organization (NATO), but anchored by friendly spacefaring states such as France, Japan, India, UAE. (OPRs: EOP, NSpC, NSC, DOS)

"The USSF is committed to its broader strategic purpose to support space as a growing element of U.S. national power …the Space Force will 'be there' wherever U.S. commercial and strategic interests and activities expand."
– USSF SPACE FUTURES WORKSHOP REPORT, 2021

LONG-TERM PAYOFF

Maintain the Cislunar Economy as an Open Economic System by bringing the Moon into Earth's economic sphere and underwriting the security of this vast theater with protective U.S. military space power, enabling the U.S. and its alliance partners to enjoy the vast wealth of the inner solar system and maintain a balance of material capabilities which favor freedom in space and on Earth. (OPRs: EOP, NSpC, NSC, USSPACECOM, USSF)

Figure 40: The Starship will be capable of delivering 100 metric tons of spacecraft, materials and commodities to low Earth orbit. Multi-orbit logistics vehicles will be ready to transport them to other destinations along the Space Superhighway (Credit: SkyCorp).

SPACE MOBILITY & LOGISTICS

"Logistics is the bridge between the economy of the Nation and the tactical operations of its combat forces. Obviously then, the logistics system must be in harmony, both with the economic system of the Nation and with the tactical concepts and environment of the combat forces."
– REAR ADMIRAL HENRY ECCLES, 1959[236]

BACKGROUND

Space transportation and logistics capabilities are critical for any expansive U.S. space future across civil, commercial, and military domains.[237] No nation has maintained a dominant position in a domain (air, land, maritime, space, and cyber) without a superior capability for movement and sustainment within that domain. Mobility and Logistics includes the transportation and transformation of both physical objects (physical logistics) and information (digital logistics). Getting hardware into orbit is one vital aspect of the infrastructure. The complement to that is retrieving data from that orbiting hardware. The digital portion of logistics includes the software and digital data standards that allow both for information transport and physical connection (discussed in more detail in the Space Information Services section).

Space Infrastructure Accelerates Space Innovation – Just as the modern highway system provides commercial opportunities within the nation, space infrastructure is the backbone of effective logistics. In the commercial domain the ability to transport to and through space reliably and cheaply will be the key determinant for the commercial viability of all other space capabilities. In the short-term this particularly drives the trajectory for personal space travel (space tourism, etc.) and supports the exploding demand for information services. In the mid- to long-term this drives the technical and fiscal feasibility of space manufacturing, Lunar resource extraction and space power. The responsiveness and flexibility provided by this infrastructure will provide numerous DOD advantages including maneuver, resilience, modernization and adaptability. The common term for this space infrastructure used at the workshop was the "Space Superhighway."

Modularity Will Transform the Industry and Space Systems -- Today's spacecraft are custom-built, highly integrated, unmaintained and disposable. No other industry that develops billion-dollar assets would use such an approach. Terrestrial industries have long benefited from modular approaches, Transitioning the space industry to modularity will have benefits both pre-launch and on orbit. Modular manufacturing lines can accommodate technical changes, correction of test anomalies, and interchangeability between multiple vendors.

Figure 41: Arkysis Port is a modular, rapidly configurable space system (Credit: Arkysis)

---

[236] Wissler, J. (2018). Logistics: The Lifeblood of Military Power. Retrieved from https://heritage.org
[237] Jehle, A. & Sowers, G. (2021). Orbital Sustainment and Space Mobility Logistics. The Space Force Journal.

In space, modular designs will allow rapid introduction of new technologies, correction of failed components, and less frequent disposal. Launching modules rather than complete spacecraft will shift the launch process from chartered to scheduled. More companies will be qualified to produce space hardware, expanding and enhancing the supply chains and industrial base. Developing the appropriate incentives to ignite the modular transformation should start now. Combined with a mesh network for in-space communication, modularity will lead to a highly resilient DOD space architecture.

Standards Reduce Commercial Risk (if done properly) – A vibrant, robust space transportation and logistics industry can support a greatly expanded set of space activities, provide a clear U.S. strategic advantage in space, and stimulate economic activity that is highly attractive to investors. Space transportation and logistics will become integrated into a single logistical system. Developing this system to support all users, civil, commercial and military, will be a powerful enhancement of national space growth and dominance. The first step in supporting this modularity-based logistical system is to select, develop, and utilize the best common interfaces for the growth of the logistics chain. These standards should create the lowest possible barrier for entry for new, small, and innovative companies, while simultaneously supporting a large and growing industry. Clear measures for the readiness and utility of interfaces need to be established; early and often flight testing is essential.

CURRENT STATE

In-Space Logistics -- A plethora of commercial launch options now exist to provide Earth-to-orbit space logistics. The first two in-space life extension servicing missions have been performed by MEV-1 and MEV-2. Several start-ups have emerged with ambitions for in-space physical logistics and refueling. At least one firm now has significant funding. NASA is helping establish commercial logistics to the Lunar surface via its Commercial Lunar Payload Services (CLPS) program. NASA's OSAM-1 mission will demonstrate in-orbit refueling, assembly and structural manufacture. DIU is leading efforts to catalyze multi-orbit logistics. Several standards have emerged for physical connections.

Plummeting Launch Costs -- the imminence of new very large launch vehicles (SpaceX Starship, Blue Origin New Glenn, ULA Vulcan) promise a decrease in launch costs to below $1000/kilogram. This will enable a blossoming of the space industry and appearance of many new applications. However, the use of such large vehicles to distribute portions of their cargoes among multiple orbits is inefficient. Just as large container ships deliver their cargoes to a single port where other transportation means (trucks, rail) deliver to final destinations, an in-space transportation system will enable the most efficient use of the coming launch vehicles.

Figure 42: Astrobotic Peregrine lander is one of many commercial logistics vehicles headed to the Moon under the NASA CLPS program (Credit: Astrobotic).

The use of Starship for rapid suborbital delivery across the globe is potentially transformational, but also drives requirements for exquisite awareness and custody of space objects along suborbital routes.

Terrestrial Supply Chain – COVID has demonstrated the volatility of resource availability for many industries, including space. Many of the raw materials and technologies required for space systems begin life outside of the U.S.. At the very beginning of our logistics chain, before any item reaches space, we need to secure the items necessary for engineering our next-generation space systems. Whether we ensure multiple sources of critical materials and technologies or we invest within our own economy to develop a surplus, not preparing for a shortage could cripple U.S. space aspirations for significant portions of time that we cannot spare.

Digital Enablers of Logistics – Interoperability between disparate systems within the space logistics chain requires more than material interoperability; it requires software and communication interoperability as well as cyber protections. The ecosystem between Earth and Cislunar space will require rapid, authenticated, and safe software and communication to ensure that no bad actors are intercepting systems and that these systems can operate as designed. Robot Operating System (ROS)[238] has continued to iterate a growth in common open software interfacing within the robotics portion of space logistics, and the commercial industry has bought in. The government must continue to help iterate within other portions of space software and communications to ensure a thriving and safe ecosystem.

In the past year, multiple instances of tipping and queuing satellites to utilize the best aspects of our commercial satellite technologies demonstrated the power of interoperable logistics within LEO. Broad coverage from large EO/IR constellations discover areas of interest using wide area analytical technologies. These broad area coverage satellites are able to provide exact coordinates to satellites with Synthetic Aperture Radar (SAR) capable of extremely high-fidelity imagery. This digital logistics chain has proven itself in rapidly tasking and imaging hot areas of interest, to include illegal fishing, smuggling, sanctions violations, military movements and construction, and work camp construction.

New Products from Space -- investments are growing rapidly into in-space manufacturing of products as diverse as human organ lattices, nanomaterials, crystals and optical products. To enable the market, an efficient downmass capability will be required, as well as a more perfect microgravity environment than the ISS provides. Space studies are also contributing to gerontology, as some effects of long-duration space flight shed light on the effects of ageing. Very large space structures manufactured in space may make significant contributions to communications and other growth markets. Lunar-sourced materials and products may improve the economics of space activities.

> *"You will not find it difficult to prove that battles, campaigns, and even wars have been won or lost primarily because of logistics."*
> – GENERAL DWIGHT D. EISENHOWER, 1945[239]

KEY ISSUES & CHALLENGES

The primary challenges to U.S. space transportation are maintaining continued growth in demand and the U.S. market position in the face of increased competition in launch services. For in-space logistics, the challenge is to expand its application beyond space exploration into commercial and military operations.

Government-Defined "Standards" Create Business Risks – Commercial space companies both need and fear standards. Standards are wonderful if they enable you to know the infrastructure

---

[238] Fong, T. (2013). ROS in Space Thoughts on Developing and Deploying ROS for Space Robotics. NTRS.
[239] Ibid, (236); Rutenberg, D. C., & Allen, J. S. (1996). *The Logistics of Waging War: American Logistics 1774-1985, Emphasizing the Development of Airpower*. Air Force Logistics Management Center.

supporting your technology and you can utilize it at a reduced cost due to the economy of scale. Interoperability standards can be extremely enabling by enabling efficient alignment, engendering competition, and fostering open systems architectures. However, no company wants the adoption of a standard to negatively impact their business by not supporting a technology they need to utilize, or worse, choose a competing technology that renders their technology obsolete. Thus, interoperability standards and frameworks must be established early and with broad industry engagement so that the government doesn't adopt a standard that the commercial industry does not adopt, creating a rift in technological development. To alleviate these challenges, the government must carefully and collaboratively forge enabling interoperability standards early and upfront, then indicate what technologies are vital for national application via investment and the purchase of services, but allow commercial consensus to drive and provide feedback on how to best deliver these requirements. Building systems that are flexible to a variety of interfaces will be optimal.

A Robust Logistics System Creates Huge Demands on Space Domain Awareness and Autonomy -- each delivery mission will require real-time support from SDA assets for safe orbit transfers and rendezvous. As the logistics system grows, control from the Earth becomes less practical and autonomous operation more essential. Yet autonomous operations pose their own challenges for safe space operations. Orbits farther from the Earth than GEO are influenced by the Moon's gravity, requiring new orbit parametrization and more intense computation. Investments in improved SDA systems, autonomous operations, astrodynamic predictors, in-space edge computing and affordable robotics will be necessary for a robust Space Superhighway.

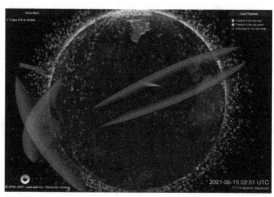

Figure 43: Leolabs is increasing its ground stations globally to provide continuous radar monitoring as a service (Credit Leolabs).

Mismatch Between R&D Funding and Space System Costs– A major concern within the commercial space community and its investors is the high cost of delivering any technology to space. Due to these increased costs, risk avoidance postures are taken and innovation suffers, because these companies and their investors are waiting on an indication from the government on what technologies the government will be selecting and purchasing. If a company requires a specific technology, such as refueling, they may not care what port is available to refuel their spacecraft, but they need to choose the correct one. By investing a substantial amount of money into a technology the government, the single largest buyer of space capabilities, indicates what technologies will likely survive the operational time of the commercial satellite. These decisions drive innovation and expand investment into U.S. space companies. Furthermore, the amount of money that the government places on a specific technology will indicate how confident the government is in that technology selection, which strongly influences the confidence of investors and developers. Small investments do not indicate confidence. Space is extremely expensive, and even as we drive the cost down, the amount of money awarded on SBIR contracts does not provide enough support to significantly support a company. These amounts are suitable for terrestrial technologies. It would make more sense to award fewer, higher dollar amount awards for SBIR Phase 2 and 3 companies in order to incentivize innovation and adequately fund the costs associated with space technology development.

KEY INFLECTION POINTS

"Between 2015 and 2025, we have an opportunity to put a fleet on another sea. And that sea is space."
– GENERAL CHARLES KRULAK, USMC, 1997

- **U.S. provides financial backing** providing early indications of interest to areas of infrastructure within the Space Industry that are not currently commercially viable, but will enable extraordinary growth opportunities and reduce foreign investment temptations

- **U.S. influences the interface standards** by purchasing the services provided by our commercial companies at the early stages, when our industrial base is most vulnerable, especially the ones that support government interfaces that are interoperable with our other investments

- **Growth in demand for broadband services continues** at its current rate of over 20%/year, driving satellite manufacturers to adopt modular, upgradeable designs to maintain market share.

- **U.S. incentivizes emerging logistics providers** to take on secondary roles that help advance new technology readiness or mitigate known risks in the space environment (e.g. pay them to remove orbital debris)

- **U.S. reduces the barrier of entry** for innovation by developing or adopting the modular infrastructure that will reward small company innovation, leveraging reduced launch costs and private investment

- **A large human presence in space** (100 people simultaneously in space? 500? 2,000?) **demands robust logistical support** which leverages the same system that supports next-generation DOD and commercial satellites

- **NASA selects an in-space assembly approach for its next large astrophysics observatory** which will require a logistics stream for delivery of components

- **Space domain awareness is extended to the entire Cislunar volume**

KEY ACTIONS & RECOMMENDATIONS

SHORT-TERM PAYOFF

DOD develops plans to place propellant sources in space or purchase propellant services with the input of NASA and commercial industry. The selected fuel source and fluid interface will help establish a logistics chain for Cislunar operations, indicate where industry should innovate, and allow for improved servicing and maneuverability of assets. (OPRs: USSF, DIU, AFRL, DARPA, SDA)

DOD increases enabling support to the space logistics industry with (1) increased values for SBIR awards; these awards will help further drive investment into innovative technologies; (2) increased R&D in DOD technologies that define the state of the art (robotics, RPO, propulsion, autonomy); (3) frequent risk-taking tests of components, systems and procedures. (OPRs: USSF, DIU, AFRL, DARPA, SDA)

DOD increases the use of commercial SDA systems to obtain the needed safety, timeliness and accuracy of logistics platforms during their operations. (OPR: USSF)

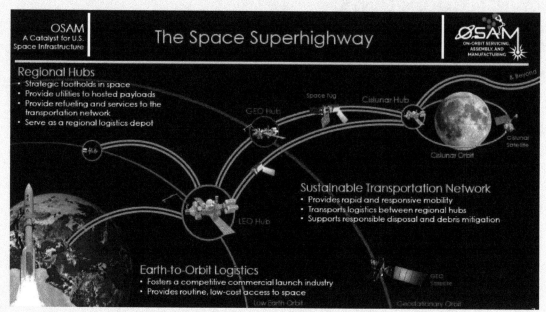

Figure 44: A hub-and-spoke concept for a future commercial space logistics infrastructure serving LEO, GEO and Cislunar customers. (Source: NASA/DOD OSAM National Initiative).

The National Space Logistics Infrastructure Concept

The Space Superhighway concept has the potential to transform American space operations from a disposable, vulnerable, aging fleet to a vibrant, dynamic, and sustainable system.[240] It promises economic impact on the order of our Interstate Highway System, the Transcontinental Railroad or other large infrastructure investments. As a general-purpose transportation and logistics system, its design would support uncrewed missions, such as GEO communications satellites and Cislunar space domain awareness spacecraft, as well as crewed missions including a sustainable Lunar presence and travel to/from Mars.

A whole-of-nation effort to establish an in-space logistics infrastructure is necessary to mutually benefit commercial, civil and national security space. Like previous infrastructure projects, a National Space Logistics architecture should highly leverage commercial capabilities. Civil space has already realized significant cost savings by utilizing commercial resupply and crew transportation services provided by a variety of companies. A space superhighway should employ a similar public-private partnership model that incentivizes private investment to develop new business in space.

Just as the Navy, Air Force, and Army benefit from civil and commercial ports, airports and roads, the U.S. Space force can leverage this commercially owned and operated infrastructure to extend its reach, improve its maneuver, lower costs, and get the technology it needs into space more quickly. Superior logistics is therefore an enabler to the Space Force missions to provide deterrence and security. The U.S. Government should be a customer and early adopter of this in-space logistics system, provider of seed funding, and developer of essential technologies, many of which are already at a high readiness level.

---

[240] Space SMART Think Tank (2021). In-space servicing, manufacturing, assembly, robotics, and transportation (SMART) initiative.

The USG engages academia in long-term challenges associated with space logistics systems, including making use of the objectives envisioned by the proposed Endless Frontier Act and the U.S. Innovation and Competitiveness Act. (OPR: NSpC, National Science Foundation, USSF, AFRL)

MID-TERM PAYOFF

DOD incorporates modular capabilities by adopting the standard containers that are proposed jointly between commercial and government development. Much like the shipping container, launch containers would make use of the total area inside a rocket. Commercial companies would be able to design to specific sizes that fit within a multitude of launch vehicles providing rapid launch or rideshare opportunities, insured containers to protect the main satellite and extra payload, and additional security for sensitive payloads. (OPRs: USSF, DIU, AFRL, DARPA, SDA)

Figure 45: An orbital propellant depot in low Earth orbit (Credit: OrbitFab)

DOD incentivizes propellant depot services or invests in capabilities for enhanced maneuver in space (LEO, MEO, GEO, XGEO, Cislunar). (OPR: USSF)

LONG-TERM PAYOFF

Fully modular logistics chain between Earth and Moon by building the logistics infrastructure iteratively and conscientiously. As each interface between spacecraft, modular components, refueling servicers, and robotic arms is widely adopted, a support system for Cislunar objectives will gradually build out, as the market is capable of supporting the next iteration of capabilities. (OPRs: USSF, DIU, AFRL, DARPA, SDA)

USG increases STEM investment in local communities with a substantial drive to educate our growing youth population. The wide human representation within America is our greatest asset, however, the representation within the space industry is not as diverse. By investing in underrepresented communities, we will enable the growth of diverse ideas that we can take to Cislunar space and Lunar orbit. (OPRs: USSF, DIU, AFRL, DARPA, SDA)

USG enables a sustainable climate future by adopting modular logistics approaches that provide more frequent and affordable monitoring of rising sea levels, illegal fishing, and oil spills. As our technologies and sensors increase in capability and our data services track environmental changes, the USG can be on the forefront of preserving the planet for future generations. (OPRs: USSF, DIU, AFRL, DARPA, SDA)

Figure 46: Virgin Galactic's VSS Unity, carrying Richard Branson and crew, accelerates at an 85 degrees nose high attitude to climb beyond the Kármán line into space on 11 July 2021 (Credit: Virgin Galactic/video capture).

SPACE POLICY & FINANCE TOOLS

"To all you kids down there, I was once a child with a dream looking up to the stars – now I'm an adult in a spaceship with lots of other wonderful adults looking down to our beautiful Earth. To the next generation of dreamers – if we can do this, just imagine what you can do."
— SIR RICHARD BRANSON, CEO Virgin Galactic, 11 July 2021[241]

BACKGROUND

Policy remains a foundational element in sustaining and strengthening the U.S. space industrial and innovation base. The well-publicized achievements of the past several years, particularly in the launch arena, have made this increasingly evident to new entrepreneurs, policy makers and the public. Recent space-flights of Jeff Bezos and Richard Branson[242] and well as the steady march of SpaceX Starship milestones have reawakened public attention in space. But the seemingly infinite possibilities for economic opportunity offered by the space domain come with the challenge of early-stage alignment and prioritization. Further, in this environment of renewed strategic competition,[243] U.S. preeminence in space is no longer assured.

The U.S. Government can play a critical role in shaping the future of the space industry and global norms. Visionary space policy can create a whole-of-nation alignment and unleash the full potential of U.S. industrial power. Long-term, forward leaning strategies supported by sound fiscal and acquisition policies can guide and decrease risk for firms and their investors. Decisive policy-action will accelerate the growth of the U.S. space industry and assure continued U.S. preeminence in space now and into the future.

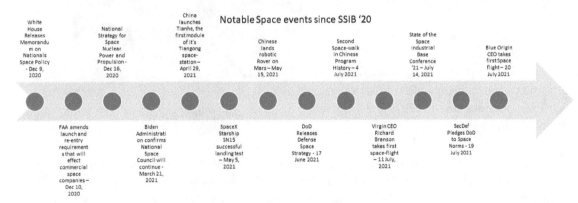

Figure 47: Notable space events since the SSIB'20 Report (Credit: DIU).

---

[241] From SpaceShipTwo on its historic suborbital flight.
[242] O'Caine, C. (2021). Billionaires Jeff Bezos and Richard Branson have now both gone to space. Here's the difference between their Blue Origin and Virgin Galactic flights. **CBS**.
[243] Broad, W. (2021). How Space Became the Next 'Great Power' Contest Between the U.S. and China. **NY TImes**.

CURRENT STATE

The U.S. space industry was initially shaking at the start of the COVID-19 pandemic but quickly bounced back, attracting record levels of private investment in 2021. According to an April 2021 report by Quilty Analytics, a space research and analytics firm, the industry has attracted nearly $17 billion of investment since 2015, a 20-fold increase from the less than $1 billion invested cumulatively over the prior decade.[244] SPACs are an increasingly common destination for space-related capital. The Biden Administration retained the National Space Council chaired by Vice President Kamala Harris, and the DOD led the government in setting tenants for responsible space operations.[245] The USSF celebrated its first birthday in December 2020 and now enters a strategic window of opportunity to set policies, regulations, and strategy to leverage the advancements in the commercial space sector and future trajectory of the space domain. The Chinese space industry continued to surge forward in 2020 with a three-fold increase in commercial space investment from 2019 levels.[246] The United States remains the global space leader, but its lead is shrinking or has been overtaken in select sectors. According to analysis from the National Geospatial Agency, China is now the world leader in several key space areas such as growth of launch[247] and overall commercial remote sensing capabilities.[248]

In the last year, the U.S. Government has laid out national and agency space strategies to include the Defense Space Strategy, National Space Strategy, and a National Strategy for Nuclear Power and Propulsion. In addition, the Federal Aviation Administration (FAA) streamlined and provided greater flexibility in space launch and relaunch regulations for commercial space companies.[249] However, while important steps forward, additional efforts to leverage USG government buying power and a more coherent long-term vision are needed.

Commercial Space Industry Attracted Record Levels of Investment – In early 2020, Intelsat, Global Eagle, and Speedcast filed for Chapter 11 bankruptcy protection, sparking funding and viability concerns for the around 450 domestic space startups established after 2015 and the broader space economy.[250] Q2 2021 was the largest quarter on record for space infrastructure investment and the historically high levels of capital raised in 2020 will likely see this trend continue in the short term.[251]

Figure 48: All-civilian Inspiration4 crew prepares for spaceflight (Credit: Inspiration4)

Commercial Space Milestones Reached – Since the beginning of 2021 the domestic space industry enjoyed several notable events including Virgin Orbit successfully reaching orbit in January, the successful landing of SpaceX Starship in May, and an Initial Public Offering by Astra in July 2021.

---

[244] Quilty Analytics (2021). Space 3.0. Turning the corner in 2021?.
[245] Secretary of Defense (2021). Tenets of Responsible Behavior in Space. Media.defense.
[246] Jones, A. (2021). China's commercial sector finds funding and direction. SpaceNews.
[247] "Leveraging the Emerging Space Economy to Meet Critical Government Needs"- Chris Quilty at SSIB '21
[248] Beames, C. (2021). Analysis: China, Europe pulling ahead of U.S. in commercial satellite imaging. SpaceNews.
[249] https://www.govinfo.gov/content/pkg/FR-2020-12-10/pdf/2020-22042.pdf.
[250] Quilty Analytics (2021). Space 3.0. Turning the corner in 2021?.
[251] Space Capital (2021). Space Investment Quarterly Q2 2021. Space Capital.

SPACs Fill a Gap in Space Investment – Q1 2021 saw the formation of 8 new SPACs focused on the commercial space industry with over $21 billion in aggregate equity value.[252] SPACs provide a 'middle ground' investment vehicle for nascent space companies that address the gap between early-stage VC investment and traditional IPOs and we expect this trend to continue; albeit at a possibly slower pace in 2022.

The National Space Council Remains an Opportunity for Positive Policy Steps – In March 2021, the Biden Administration announced that the National Space Council (NSpC) would continue. In April, it was announced that Vice President Harris would lead the council and intended to steer the agenda toward issues such as climate change, cyber security, and STEM education.[253] With key cabinet officials as members of this group, and a diverse group of space-professionals as members of the National Space Council Users Advisory Group, this body is poised to become a powerful forum for the development of a more complete future vision for the United States efforts once it convenes under Vice President Harris's leadership.

Intelligence Community Commercial Space Council Established – It is also encouraging to see the Intelligence Community establish the IC Commercial Space Council as a means to address IC-specific interagency issues and priorities and to assist in the policy formation for this rapidly changing sector. The establishment of a Space Information Sharing and Analysis Center (S-ISAC) is another cross-domain, interagency mechanism that has been established similar to those in other domains.

DOD Makes First Step Toward Leadership in Establishing Space-domain Norms – In a July 7 memo, Secretary of Defense Lloyd Austin pledged the DOD to abide by a set of norms and established five "Tenets of Responsible Behavior."[254] The memo drew from 1967 Outer Space Treaty language concerning operating with 'due regard' and addressed the need to limit the creation of 'long-lived debris' and harmful interference. It tasked the Undersecretary for Policy to lead the implementation of these tenets in the DOD as well as the rest of the U.S. Government and specifically mentioned 'International Relations.' This is a positive step toward a key 'enabling inflection point' identified by the SSIB'2020 report.[255]

"One can imagine a self-reinforcing virtuous cycle of development that would support the space economy. But one can also reasonably doubt that such an ideal path will be realized easily or without some nudges along the way. Limits on or asymmetries of information, the high level of risk inherent in space and the challenges of capturing surplus from such complementarities will make it difficult to move forward on the most efficient path—or even to move forward at all."
– MATTHEW WEINZIERL, Harvard Business School Professor, 2018.[256]

---

[252] Space Capital (2021). Space Investment Quarterly Q1 2021. Space Capital.
[253] Foust, J. (2021). Harris to place 'personal stamp' on National Space Council. Space News.
[254] Secretary of Defense (2021). Tenets of Responsible Behavior in Space. Media.defense.
[255] USSF-DIU-AFRL (2020). State of the Space Industrial Base 2020 Report. AFRL. See Appendix B.
[256] Cahan, B. & Sadat, M. (2021). U.S. Space Policies for the New Space Age: Competing on the Final Economic Frontier. NSNM. footnote 88 on page 50 in

KEY ISSUES & CHALLENGES

Aligning Government Efforts – Sustainment of U.S. space leadership will require the collective efforts of dozens of departments and federal agencies. Sound policy is needed to signal space as a priority for agency resources and initiatives, and to direct interagency cooperation. A multi-decade outlook is needed to shape long-lead efforts such as new-space technology and the growth of the robust STEM workforce needed to power the future space economy.

Space Narrative Must Align to Administration Priorities and Broaden Public Interest – The space industry and government space experts must do more to show how advancements in space will support the Biden Administration's stated goals for reducing greenhouse gas emissions, job-creation and infrastructure investment. Future technologies such as Space-based Solar Power generation will significantly reduce reliance on fossil-fuel based energy production. Rare-earth minerals mined from celestial bodies will enable cost-reductions for the electric economy. The growth of new space-related industries will create new and *new-kinds* of job-opportunities, many of which may replace those absorbed by automation.[257] The space industry is projected to triple to $1.4 trillion within a decade.[258] A healthy space infrastructure supported by a STEM workforce must be put in place to capture a dominant share of this future economic growth.

Figure 50: Global competition for space jobs: Astra rocket factory in Alameda CA (*left*/Credit: Astra), and China's first smart manufacturing plant for satellites in Wuhan, Hubei province (*right*/Credit Chang Jiang Daily)

---

[257] McKinsey Global Institute (2017). Jobs Lost, Jobs Gained: Workforce Transitions In A Time Of Automation.
[258] Sheetz. M. (2020). Bank of America expects the space industry to triple to a $1.4 trillion market within a decade. CNBC.

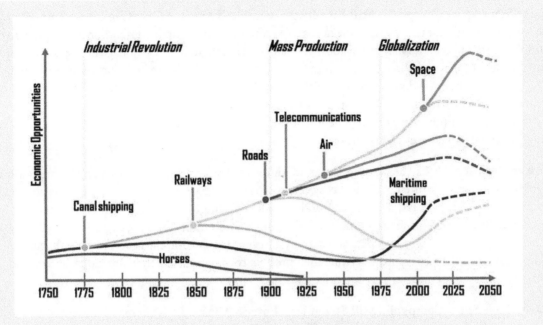

Figure 49: The introduction of new transportation modalities and their impact on new economic opportunities (Source: HOP and Associates)[259]

Economic Growth Follows New Transportation Modalities

Throughout human history, the greatest advancements in economic opportunity can be intrinsically linked to the introduction of new transportation modalities that have forever changed the economic and military influence in state affairs. History also reveals that such opportunities are fleeting. In other words, there is a significant *first mover advantage* for those who recognize and establish early entrance and leadership in new and emerging transportation markets. We can see this occurring within the commercial launch sector today where early entrants such as SpaceX and Rocket Lab USA hold a significant share of their respective markets while a large number of late entrants struggle to establish a foothold. The market has rapidly reached a saturation point where only a few firms will ultimately prevail.

In 1588, the Netherlands launched an innovative era of scientific, economic and military advancement that we refer to today as the Dutch Golden Age. This era spanned more than a century and led to the Netherlands becoming the foremost maritime and economic power in the world.[260] In 1602, the Dutch East India Corporation was founded, becoming the first multinational corporation. An economic and military power shift of this magnitude is possible with increased access to space in the next few decades.

China is committed to surpassing the United States as a great scientific, economic and military power in space. This can only happen if the U.S. cedes its *first mover advantage* in shaping the space economy.

---

[259] Adapted from HOP Associates (2005) "Time, mobility and economic growth"
[260] Ten Raa, T., Mohnen, P., van Zanden, J. L., & van Leeuwen, B. (2009). Invention, Entrepreneurship and Prosperity: The Dutch Golden Age. Available at SSRN 1528208.

Early-Stage Space Companies Still Face Investment Headwinds – The VC industry continues to be a major contributor to growth of the new space economy, but the upward trends in capital deployed and the flurry of headlines may overshadow fundamental misalignment. Early-stage new space ventures require more capital over a longer horizon to fund both the hardware-based product developments and the 'long-game' business development activities needed to court government customers. These new firms are unlikely to reward their investors with returns in the 2-5 year time horizon often seen from software-based startups. This misalignment will exacerbate the effect of future economic downturns as this 'impatient-capital' seeks safer harbors.

Private Industry Lacks the Needed Demand Signals – The critical technologies and future capacity needed to power the future U.S. Space ecosystem may take 10-15 years to develop. Industry is poised to address future needs but requires a clearer understanding of the government's requirements and priorities. By defining future needs for things like launch (and return) capacity, fuel and propellant requirements, Earth observation data and communications bandwidth, the government will help existing and future space companies refine their product roadmaps, focus their limited resources, and articulate market opportunities to potential investors.[261]

Early-Stage Space Companies Still Struggle to Enter a 'Closed Ecosystem' – New entrants seeking government business struggle with a paradox. They seek support from financiers who want the confidence of contract awards which the contracting agencies won't award to firms without healthy balance sheets (enabled by stable financing). The chasm between SBIR awards and successful transition remains vast. The security clearance process remains a barrier to new firms' product development and business development in a case of 'the rich getting richer.'

The U.S. Will Need a STEM-Powered Workforce to Compete in Space – Success in space will require the development of new technologies and will test existing ones. The development, production and operation of these future enablers will require a highly educated workforce across all Science, Technology, Engineering and Mathematics (STEM) disciplines not only in design but skilled labor. Actions must be taken today to identify key skill sets, develop education programs and training pipelines, and ramp 'production' in time to meet future human capital demands. The people who will form the foundation of our space-workforce in 2030 are choosing their career-paths *today*.

Figure 51: Space Camp, like Scouting, delivers hands-on STEM experiences to young people (Credit: Space Camp)

> *"If we do not make the strong effort now, the time will soon be reached when the margin of control over space and over men's minds through space accomplishments will have swung so far on the Russian side that we will not be able to catch up, let alone assume leadership."*
> – LYNDON B. JOHNSON, Vice President, 1961[262]

---

[261] See Appendix C.
[262] Johnson, L.B. (1961). Evaluation of Space Program (VP, Memorandum for the President). NASA.

KEY INFLECTION POINTS

- **U.S. outlines a National 'North Star' vision for space** which articulates how the U.S. will sustain its space superiority through at least 2050. This vision enables key government agencies to align their activities in support of U.S. space superiority, and subsequently develop requirements for industry. Industry can begin to make investments in research and development, corporate infrastructure and human capital to support meeting future demand.

- **U.S. remains and expands as the 'partner of choice' for emerging space powers** such as Brazil, India, and Japan, and participates in establishing internationally recognized standards and practices.

- **Sustained human presence on the Moon by a space-power** would accelerate that nation toward key space capabilities such as manufacturing-at-scale, resource extraction, and basing for activities in Cislunar space and beyond.

- **China becomes the leader** by out-competing the U.S. in providing lower-cost, higher-capability space services to U.S. allies and partners.

- **The United States fills a vacuum in space leadership** brought on by a global crisis. There may have been opportunities missed during the COVID-19 pandemic.

Figure 52: More than 200 Chinese students attend a Q&A session on 3 September 2021 with Chinese astronauts, space engineers and experts. (Credit: Xinhua)

> *"The exploration of space will go ahead, whether we join in it or not and it is one of the great adventures of all time and no nation which expects to be the leader of other nations can expect to stay behind in this race for space."*
> – JOHN F. KENNEDY, President, 1962[263]

KEY ACTIONS & RECOMMENDATIONS

SHORT-TERM PAYOFF

Recognize space or key space capabilities as U.S. 'Critical Infrastructure' and establish funding and incentives commensurate with its importance to U.S. security and prosperity. Key enablers would be a Space Commodities Exchange (see below) and attractive financial incentives for space-investment (see below). (OPRs: NSpC, NEC, OSTP, DHS)

Establish a U.S.-based Space Commodities Exchange, echoing a recommendation SSIB '21. The NSpC and National Economic Council (NEC) should direct the Secretary of Commerce, the Chair of the Commodity Futures Trading Commission (CFTC) to identify the necessary steps to create a Space Commodities Exchange.[264] This would enable a granular view of pricing and market demand and enable planning and valuations by the private sector. (OPRs: NSpC, NEC, OTMP, DOC, SBA, CFTC)

Adopt a holistic readiness level framework such as the System Readiness Level Metric[265] in order to translate TRL, MRL, IRL and business readiness into a metric for commercialization that can be understood and considered by the capital markets. (OPRs: NIST, SEC, NASA)

Adopt a scheduled National Security Space Launch strategy as opposed to the current 'charter' scheduling method in order to more clearly define future requirements for vehicles, propellent and other critical long-lead items. (OPRs: OSTP, NSC, OMB)

Create a Chief Economist within the Department of Defense to advise OMB and the Dept of the Treasury on defense technology budgeting and financing considerations. This expert would report on the potential economic benefits of defense technology, as well as the potential economic impact to lost capabilities. They would make recommendations on funding and financing instruments. (OPRs: DOD)

> *"We estimate that GPS has generated roughly $1.4 trillion in economic benefits (2017 dollars) since it was made available for civilian and commercial use in the 1980s … We estimate that the loss of GPS service would have a $1 billion per-day impact."* – NIST, 2017[266]

---

[263] Kennedy, J.F. (1962). Text of President John F. Kennedy's Moon Speech at Rice University. NASA.
[264] Cahan, B. & Sadat, M. (2021). U.S. Space Policies for the New Space Age: Competing on the Final Economic Frontier. NSNM. pages 76 – 80.
[265] Cahan, B. and Ross, S. (2021). Lean technology development baseline using readiness level metrics. Presentation at SSIB'21.
[266] NIST (2017). Economic Benefits of the Global Positioning System (GPS).

Figure 53: GPS III SV04 is encapsulated within a SpaceX payload fairing on Sept. 21 in preparation for its launch on 29 Sep 2021 from Cape Canaveral Air Force Station, Florida (Credit: Space Launch Delta 45)

MID-TERM PAYOFF

Address structural barriers to new entrant competition by modifying contract selection criteria to allow small firms with fewer financial resources to compete, and by creating a mechanism for firms without clearance to begin to address critical government challenges while preserving national security. (OPRs: USSF, NASA, SBA)

Establish a Strategic Propellent Reserve similar to the Strategic Petroleum Reserve (SPR) in order to provide initial demand signal to industry and ensure the future availability of this fundamental and critical enabling resource. (OPRs: DOE, USSF, NASA)

Incorporate financial incentives for investors to enter the space sector with 'patient capital.' These may include favorable tax treatment of (sufficiently long-term) capital gains e.g., expand or extend Qualified Small Business (QSB) Stock Tax Benefits. This might also include corporate tax deductions for space-related projects or space-related capital expenditure. (OPRs: Treasury)

LONG-TERM PAYOFF

Establish a 'STEM ROTC' or similar scholarship program that targets undergraduate students in technical majors who agree to be placed in relevant positions for a period following graduation. This will strongly signal the importance of the technical disciplines to our future economy and narrow the cultural gap between government, industry, and academia. (OPRs: DOD, DoEd, DoE, NSF)

Figure 54: A cluster of three Hawkeye 360 small satellites flying in formation (Credit: Hawkeye 360)

SPACE INFORMATION SERVICES AND THE HYBRID SPACE ARCHITECTURE

"The secret of war lies in the communications."
– NAPOLEON BONAPARTE

"If you communicate effectively, you need not fear the result of a hundred battles."
– SUN TZU

BACKGROUND

Despite a strong growth in the space information services industrial base in recent years, the Department of Defense has still not embraced the available diversity of competition. The Defense Science Board recommends that DOD should invest about 3.4 percent of its budget in science and technology, keeping pace with high-tech industries that invest about 3.4 percent of sales revenue in research.[267] DOD's FY2021 and FY2020 budget requests for science and technology were $14.1 billion, or approximately 2 percent of total spending—well below the 3.4 percent target.[268] Relatedly, funding within the DOD's RDT&E budget for system development and demonstration, which serves as a bridge for getting science and technology prototypes into the force, fell to 15 percent of total research and development contract obligations, far below the historical average of 27 percent.[269] Startups can currently pursue commercial readiness through offices such as DIU, AFWERX and SpaceWERX, but often are required to partner with traditional primes to get on Programs of Record due to limitations in traditional funding vehicles. Despite these agile commercial services surpassing even some exquisite systems, government culture is still to become the dominant customer, then progressively enforce requirements until they become exquisite systems dependent on government contracts. With the surfeit of data and information services now available the government must pivot to purchasing information as a service, and allow companies providing them to remain commercially sustainable by selling in foreign markets.

GEOINT Singularity - The predicted GEOINT Singularity[270] argues that the combination of remote sensing, AI, and increased connectivity will provide the average citizen on the ground a "tremendous wealth of information, insight, and intelligence." However, the NGA analysis of commercial markets ranks the U.S. significantly behind China and the EU in their commercial GEOINT "Olympics."[271]

Ubiquitous space information services should be pursued with a sense of urgency, because U.S.' competitors are undercutting the U.S. on service cost. Any capabilities that the USA tries to withhold

---

[267] Sargent, J. (2018) *Department of Defense Research, Development, Test, and Evaluation (RDT&E): Appropriations Structure*, CRS Report No. R44711. Congressional Research Service, .
[268] Michèle Flournoy, M. & and Chefitz, G. (2020). Sharpening the U.S. Military's Edge: Critical Steps for the Next Administration. CNAS.
[269] McCormick, R. (2019). Defense Acquisition Trends 2019: Topline DOD Trends. CSIS.
[270] Koller, J (2019). The Future Of Ubiquitous, Realtime Intelligence: A Geoint Singularity. Aerospace.
[271] Beames, C. (2021). Analysis: China, Europe pulling ahead of U.S. in commercial satellite imaging. SpaceNews.

from the international sphere have rapidly been fielded by competitors, either as indigenous capabilities or by purchasing underutilized U.S. industry. It was highlighted throughout the conference that China is gaining political capital with their international Belt and Road Initiative. The U.S. should compete by making unclassified commercial space services available to allies and partners.

Hybrid Space Architecture - General Jay Raymond (Chief of Space Operations, USSF) has repeatedly and very publicly stated that a Hybrid Space Architecture represents the future force design of the USSF. This architecture will combine new space capabilities with traditional government systems. It integrates commercial, allied, and government satellites large and small from Low Earth Orbit to deep space. Integrated systems provide greater resilience and far better support to traditionally under-served users.

CURRENT STATE

Space information services have continued their renaissance over the past year despite the COVID-19 pandemic. Proliferated Low Earth Orbit constellations are dominant, with plans for expansion to Middle Earth Orbit, Geostationary Orbit and beyond. This infrastructure is a critical enabler for a Hybrid Space Architecture. The working group identified the progress of the space industrial base, on the back of private investment, as 'GREEN.'

It was repeatedly noted during the conference that while the halls of government have consistently and universally identified a Hybrid Space Architecture as essential for national security, the financial mechanisms to embrace commercial space infrastructure have not yet been found, and cultural resistance to change is proving a large impediment. The working group identified the U.S. Government's contribution to space information services as 'YELLOW-RED.'

Space-Based Communications, Internet and Broadcasting - Commercial communications networks in particular have dramatically expanded their proliferated Low Earth Orbit constellations, while showing a recent trend towards optical inter-satellite links to decrease data latency. Since the previous SSIB conference SpaceX has expanded its Starlink constellation from 360 v1.0 satellites to 1740 v1.0 satellites,[272] including 13 in polar orbit. The initial tranche of 60 v0.9 Starlink satellites de-orbited this year. SDA contracted 28 satellites for Tranche 0 of their Transport Layer, and announced Tranche 1 will include 150 satellites.[273]

Figure 55: Optical inter-satellite links provide secure, low latency communications (Credit: General Atomics)

Amazon's Kuiper, Telesat's Lightspeed, and CASC's Hongyan constellations continue planning launches. OneWeb declared bankruptcy in March 2020 but emerged later that year with financing from the UK government and Bharti Global and have since launched 180 satellites. XingYun emerged as a space-based Internet-of-Things network provider under China Aerospace Science and Industry Corporation (CASIC), launching their first two satellites in May 2020. Swarm launched 106 additional

---

[272] Clark, S. (2021). SpaceX is about to begin launching the next series of Starlink satellites. Spaceflight Now.
[273] Erwin, S. (2021). DOD space agency to award multiple contracts for up to 150 satellites. SpaceNews.

SpaceBEE IoT satellites and Canadian startup Kepler Communications launched 12 IoT satellites since the last SSIB conference.

Geosynchronous information services remain strong, especially within state-sponsored enterprises, with new satellites launched by APStar, ROK Army, RSCC Ekspress, BSAT, IntelSat, China Satcom, JAXA, Sirius-XM, ISRO, and Türksat.

Space-Based Position, Navigation and Timing - Government-owned solutions for space-based position, navigation and timing remain robust. The BeiDou-3 constellation was completed with the launch of BeiDou-3 G3Q in June 2020,[274] and the GPS Block III rollout has continued with 3 more satellites launched in the last year. Meanwhile commercial services remain constrained by the free, government-provided services, but have found some niche augmentation services.

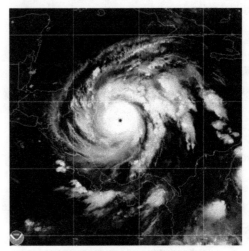

Figure 56: Hurricane Iota in the North Atlantic Ocean (Credit: NOAA)

Space-Based Weather Sensing - Government acquisition of commercial weather monitoring and prediction has been stifled by international legislation, especially World Meteorological Organization (WMO) Resolution 40,[275] requiring governments to freely share all meteorological data (thus undermining the business plan of any commercial provider). The new U.S. Government administration indicated that climate change will be a high priority over the coming years, signaling a potentially fertile market for commercial sensing despite its own related WMO Resolution on sharing.[276] As the effects of climate change are realized globally the market for related information services is likely to be broad and international, but a solution is required to incentivize commercial data collection and sharing. Traditional government owned weather monitoring relies primarily on optical and infrared sensors. Services such as radio occultation and reflectometry may provide additional inputs to meteorological models. Many smallsats have been launched in the past year to perform these experiments but remain in the technology demonstration phase primarily for lack of commercial customers and mature models capable of ingesting new data types.

Active and Passive Earth and Space Observing - Space situational awareness remains dominated nationally and internationally by defense, civil and intelligence systems, but there was slow growth in commercial assets. LeoLabs commissioned their fourth radar, based in Costa Rica, in April 2021. They also raised $65 million in their Series B to grow their ground-based network and train their analytics. Canadian company NorthStar Earth and Space has contracted Thales Alenia to build a constellation of three sun-synchronous satellites to monitor the near-Earth environment, providing space situational awareness.

---

[274] Cozzens, T. (2020). China completes BeiDou-3 worldwide navigation constellation. GPS World.
[275] World Meteorological Organisation (1995). Resolution 40: WMO policy and practice for the exchange of meteorological and related data and products including guidelines on relationships in commercial meteorological activities. WMO.
[276] World Meteorological Organisation (2015). Resolution 60: WMO policy for the international exchange of climate data and products to support the implementation of the Global Framework for Climate Services. WMO.

High-resolution imagery and radiofrequency (RF) signal collection remains the domain of exquisite defense and intelligence systems, but proliferated Low Earth Orbit commercial constellations continue to grow, backed by robust private sector investment, providing abundant low to medium resolution imagery and RF signal geolocation information. Growth is now constrained primarily by downlink windows. Government financing has not yet found streamlined pathways to acquire this abundant data, and defense and civil agencies are still learning how to ingest this varied data.

Figure 57: Leolabs Costa Rica site (Credit: Leolabs)

KEY ISSUES & CHALLENGES

Warfighters, including effectively all COCOMs, have made strident demands for more releasable data. Furthermore, General Raymond has repeatedly called for a hybrid space architecture to provide resilience to U.S. space assets. However, these goals are facing significant systemic and structural roadblocks:

Acquisition and Contracting - Federal acquisition regulations are ill suited to buying data as a service.

Security Clearances - Commercial companies have trouble getting facilities clearances and even maintaining personnel clearances, impeding the necessary conversations surrounding government ISR needs.

Responsible Office - The NRO CSPO, currently a central procurement office for commercial imagery data and services, is limited in its ability to serve COCOM interests. Despite availability and utility of commercial data, the CSPO is only buying limited RF geolocation, SAR, and other data types important to the intelligence community for GEOINT applications and is not able to purchase from allied nations. These constraints have limited the tactical utility of NRO data to the warfighter for other applications such as situational awareness. The slow pipeline for tasking, collection, processing, exploitation and dissemination required for intelligence applications have further degraded the utility of NRO services in a tactical environment.

Data Types - The hybrid space architecture needs more than ISR. It needs PNT, weather, communications, and SSA. Since USSF needs a mechanism to buy these services anyway, ISR should be included in their contracting options. NGA will provide expert analysis of ISR products on request.

Exportability. The slow export classification and review process, a broken appeal process, and risk that new commercial data services be classified and controlled as a "munition," are constraining U.S. companies as they build greater cooperative pathways with our allies and partners. Although many of our partners currently prefer U.S. commercial capabilities, U.S. companies are today disadvantaged in the competitive global marketplace by misdirected regulation. They are at risk of losing market share and influence to foreign companies who do not have similar constraints, in particular near-peer suppliers, who are racing to sell their products in the international market.

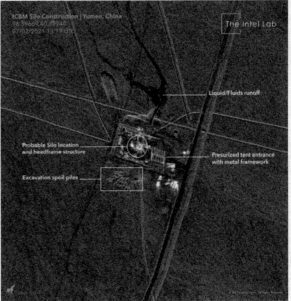

Figure 58: TOP LEFT: Planetscope low resolution image of Yumen, China, taken on 9 June 2021 illustrating multiple silo sites constructed since the beginning of the year. BOTTOM LEFT: Planet high resolution image of one ICBM silo construction site taken on 28 June 2021. RIGHT: Synthetic Aperture Radar (SAR) image of the same silo complex taken on 2 July 2021 revealing features that prove the construction is for nuclear ICBMs rather than a wind farm as claimed by the PRC. (Credit: Planet & Capella Space)

Student discovers 120 ICBM silos under construction in China using commercial remote sensing services

Decker Eveleth, an undergraduate student at Reed College in Portland, Oregon, has a passion for nuclear non-proliferation. In addition to pursuing a degree in political science, Decker collaborates with researchers at the James Martin Center for Nonproliferation Studies in Monterey, California. Earlier this summer, Decker discovered 120 intercontinental ballistic missile (ICBM) silos under construction in Yumen, China. He did so by reviewing low resolution commercial imagery of the region performing a time-phased study since the beginning of this year.[277] To confirm his findings, researchers were able to task high resolution (<1 meter) Planet Skysat and Capella Synthetic Aperture Radar (SAR) imagery shown above.

Commercial remote sensing and advanced analytics services have revolutionized our ability to observe the planet, detecting anomalous behavior such as illegal fishing or physical changes such as the destruction of property following devastating wildfires and hurricanes. Equally as important, commercial imagery is UNCLASSIFIED which means it can be readily shared with friends, allies, regional partners and non-government organizations.

Unfettered access to commercial imagery and analytics collected at the limit of continuous satellite coverage (as provided by commercial mega-constellations) is critical to increasing situational awareness and improving decision making during peacetime, contingency and war.

---

[277] Vance, A. (2021). The Undergraduate who found a nuclear arsenal. BLOG: Valley of Heart's Delight.

KEY INFLECTION POINTS

- **Establishment of a USSF commercial space services acquisition office** with new authority to contract and finance rapid, tactical tasking, collection, AI/ML processing, exploitation, and dissemination of commercial space-based data to the warfighter.

- **Adoption of hybrid space architecture standards** as exquisite MEO and GEO satellites are upgraded with modular and servicing capabilities allowing Hybrid Space Architecture (HSA) communications modules to be attached, enabling diverse secure networking with proliferated smallsat constellations.

- **Maturation of concepts of operations** that can use diverse commercial space data. There are well defined CONOPS using well defined traditional siloed information services. Commercial services are still a mostly unknown entity, and therefore CONOPS have not been developed to fully exploit them.

- **Improvements in cloud analytics and edge processing** generate more actionable intelligence from the multitude of raw data currently available.

- **Long term leadership and funding** will allow long term planning. Space is inherently a long-term endeavor. Consistent, sustained progress needs leadership positions and funding to reflect that, such as those recommended by Section 202 of HR4945 the "American Space Renaissance Act."[278]

- **The GEOINT singularity** will occur as sensors, on-orbit computation, AI/ML analytics and ubiquitous connectivity converge.

KEY ACTIONS & RECOMMENDATIONS

SHORT-TERM PAYOFF

Fund agencies to start a Hybrid Space Architecture program including an additional $50 million per year to AFRL/RV for systems development and an additional $100 million per year for NRO CSPO and NGA to facilitate adoption of commercial data and services. (OPRs: Congress)

Declare security clearances a national asset and create a security clearance management account to maintain clearances for personnel between jobs or while awaiting facility clearances, and potentially to sponsor new clearances for individuals and companies essential to the space industrial base. (OPRs: USSF)

MID-TERM PAYOFF

Expand the mission of the USSF Space Systems Command (SSC) Commercial Satellite Communication Office (CSCO) to contract for all varieties of commercial space data, products and services to include communications, PNT, ISR, SSA and analytics. CSCO will publish data needs with associated funding levels to send clear messages to industry and stimulate private investment. (OPRs: Congress, USSF)

Create a $1 billion Working Capital Fund managed by CSCO, ramping up to this level over 3 years. DOD users would purchase commercially derived data, products, and services from CSCO Working

---

[278] U.S. Congress (2015). H.R.4945 - American Space Renaissance Act. Congress.gov.

Capital Fund, who would in turn purchase these from industry. COCOMs would manage tasking within their AOR. The AFRL Global Unified Environment can be matured and expanded to help automate this process. (OPRs: Congress, USSF, COCOMs)

DOD provides and invests in high-bandwidth networked space communications capable of allowing secure, authenticated, interoperable communication between satellites. This area of interest can be additionally supported by commercial development, but the government needs to specify how these technologies will communicate in a standard way. (OPRs: USSF, DIU, AFRL, DARPA, SDA)[279]

LONG-TERM PAYOFF

Increase funding to CSCO to develop integrated systems and tools and to onboard new entrants to the hybrid space architecture via DIU and other agile onboarding mechanisms. As the Administration undertakes its review and update of national space policies, as well as defense and intelligence strategies, there is an opportunity to realize longstanding policy to "rely to the maximum practical extent" on commercial capabilities. This would start with taking a fresh look at the role that commercial space plays in hybrid architectures. Commercial space can also contribute to experimentation with new operating concepts, new distribution networks, and new pathways to share information with allies and partners, as well as being a prime candidate for exploring new acquisition models that are better suited to the pace of commercial technology innovation. (OPRs: Congress, USSF, DIU)

Create and partially fund a space internet consortium of key enabling companies to set standards and protocols, develop reference technologies, and coordinate efforts to create a space internet that allows secure transmission and sharing of data across commercial and government systems in space and on the ground. (OPRs: Congress, USSF, industry)

---

[279] Recommendation originated in Space Logistics and Mobility Working Group

Figure 59: Space is finally open to all: young and old, rich and poor and from all walks of life thanks to American ingenuity and entreneurship. Photos: Sir Richard Branson aboard SpaceShipTwo (*top left*/Credit: Virgin Galactic); Jeff Bezos aboard the New Shepard capsule (*top right*/Credit: Blue Origin); and the crew of Inspiration4: Jared Isaacson, Chris Sembroski, Sian Proctor and Hayley Arceneaux aboard the SpaceX Dragon capsule (*bottom left to right*/Credit: Inspiration4)

EVOLVING SPACE OPERATIONAL AND SUPPORT CONCEPTS

"The goal isn't just scientific exploration... it's also about extending the range of human habitat out from Earth into the solar system as we go forward in time... In the long run a single-planet species will not survive... There will be another mass-extinction event. If we humans want to survive for hundreds of thousands or millions of years, we must ultimately populate other planets.... I'm talking about that one day, I don't know when that day is, but there will be more human beings who live off the Earth than on it." -- HON. MICHAEL D. GRIFFIN, 2005[280]

BACKGROUND

The Cusp of Dramatic Change - The United States is on the cusp of a dramatic change in space access for its citizens driven by a revolution in the nation's spaceflight capabilities and in public and private sector attitudes on the 'how, who and why' of human presence in space. The United States can progress from a nation with a robust space program to a true spacefaring nation.

Changing Public-Private Roles - From its beginnings during the Cold War, human presence in space has been dominated by the U.S. and other governments. Systems for human access to space were built for and controlled by governments. This is changing. While the government is still a prime investor in human space flight for exploration, government investment is now coupled with, and augmented by, ongoing and planned large-scale private sector investments. This trend is enhanced by a growing recognition on the part of lawmakers, NASA and DOD that a key element of a sustainable path for long-term U.S. economic and strategic dominance is a catalyzed, vibrant and growing U.S. private industrial base. The driving force for this private sector investment is both economic and aspirational with the goal of enabling human expansion into the solar system and for space to become a domain of permanent occupation.

Enabling and Securing Commercial Human Ventures - *The central finding for this area is that the U.S. has an essential national interest in enabling and encouraging U.S. citizens to venture into space*, beginning with visits, and leading to the establishing long-term presence and permanent human communities. Further, DOD in its role of securing space in support of national power must leverage these civil and commercial efforts wherever possible to enhance their mission and must provide a stabilizing and protective presence.

Human Space Activities and Strategic Relationships – The expansion of the human species into the solar system is inevitable, but it is not inevitable that the U.S. plays a leading role in it. Other nations will look to partner with whatever nation is making the boldest, safest, and most economically productive forays into space. While some would characterize human spaceflight as a luxury, it should rather be seen as an essential element in the overall strategic goals of the United States.

---

[280] Washington Post (2005). NASA's Griffin: 'Humans Will Colonize the Solar System'. The Washington Post.

CURRENT STATE

The U.S. is On the Threshold of a New Era in Human Presence in Space - Almost ten years after the retirement of the Space Shuttle, Space Exploration Technologies (SpaceX) launched on 30 May their DM-2 mission carrying two NASA astronauts to the International Space Station (ISS) for a 3-month mission. This flight is the culmination of the NASA Commercial Crew program which helped create SpaceX's Dragon 2 and Boeing's Starliner for the purpose of transporting NASA astronauts to ISS and back. Equally important, this Commercial Crew Program is a major departure from traditional NASA practices with Boeing and SpaceX owning and operating their vehicles with NASA as a customer; hopefully one of many future customers. In the next several years a number of new, human-capable, launch systems are scheduled for operations including the ULA Vulcan/Sierra Nevada Dream Chaser, Blue Origin New Glenn and the SpaceX Super Heavy/Starship. They offer the promise of increasing launch-mass to orbit, to the Moon or Mars in excess of 100 metric tons and to reduce launch cost below $100/kg.[281] In addition, both Blue Origin and Virgin Galactic will soon commence suborbital flights for personal space travel. Longer term both NASA[282] and DARPA (via its DRACO[283] program) are pursuing nuclear thermal propulsion.[284]

Government-Industry Teaming is Enabling Lunar Access and Logistics - NASA is aggressively pursuing the Artemis program to return humans to the Moon by 2024. Supporting that goal NASA is executing the Space Launch System/Orion heavy launcher program and has awarded contracts for the Human Landing System[285] to take astronauts to and from the Lunar surface with stays of up to two weeks. The current awardee is SpaceX with their Lunar-customized Starship. SpaceX has also won a contract for logistical supply of the Lunar Gateway. NASA is developing the capabilities for a sustained presence on the Moon starting in 2028 which will require more intensive logistical support. Earlier, NASA established the Commercial Lunar Payload Services (CLPS)[286] program that will enable robotic commercial landers to help with advance survey and prepositioning, and to spur private access to the Moon. Both private industry and government are investing in robotic systems that will increase the efficiency of human operations in space and on planetary surfaces, as well as being the key technology for a robust and affordable Cislunar logistics system.

Figure 60: 3D printed habitable structures using Lunar regolith as the primary building material (Credit: Icon)

---

[281] SpaceX. (2020). Starship Users Guide.
[282] NASA (2021). NASA Announces Nuclear Thermal Propulsion Reactor Concept Awards. NASA.
[283] DARPA (2020). Demonstration Rocket for Agile Cislunar Operations (DRACO). DARPA.
[284] Gilbert, A. (2021). Enhancing Military and Commercial Spacepower through Nuclear. The Space Force Journal.
[285] NASA (2021). As Artemis Moves Forward, NASA Picks SpaceX to Land Next Americans on Moon. NASA.
[286] NASA (2021). Commercial Lunar Payload Services Overview. NASA.

Government-Industry Teaming is Enabling LEO Industrialization - Last year NASA unveiled a new ISS commercialization plan to encourage research and manufacturing and significantly, to allow two commercial astronaut visits to Station of up to 30 days each per year using Dragon 2 or Starliner.[287] Also, Axiom Space won an award from NASA to use an ISS berthing node to host a commercial module offering commercial services with a 2024 launch date. Axiom recently announced that they had signed a contract with SpaceX to take 3 passengers (each paying $55 million) and an Axiom commander on a 10-day trip to the ISS in the second half of 2021.[288] Space Adventures, which had previously brokered trips for 7 private astronauts to the ISS on the Russian Soyuz, has also contracted with SpaceX to perform a free-flyer mission that does not dock with ISS that could also send four private astronauts on a five-day orbital flight in late 2021.

Commercial Ventures are Supplementing Government Collection of Space Domain Awareness (SDA) Data – With increased human space travel will come the need for more timely and thorough space traffic awareness and management. Multiple companies are developing robust sensor networks that will provide greatly improved SDA in low Earth and geosynchronous orbits. Recent results indicate that space objects can be detected and tracked much deeper into Cislunar space,[289] which will be essential for the safety of coming Lunar missions.

KEY ISSUES & CHALLENGES

Expanding Access - While these technical advances are impressive the key issue is how to promote a virtuous cycle for commercial expansion of human presence in space beyond what is done for exploration and the fledgling capabilities for personal space travel offered by SpaceX, Virgin Galactic and Blue Origin. The challenge is to transition from personal space travel as a domain for the rich or a limited domain for civil human presence for exploration to one that supports multiple commercial activities in and through space and leads to a sustained presence of humans in space and on other celestial bodies and planets. This transition has three elements:

- **Make it Safer** – Drive towards space travel being as safe as air travel is today.

- **Make it Desirable** – Drive towards making travel to/from and habitation in space easy and enjoyable.

- **Make it Affordable** – Drive down costs towards a price point <$100/kg that will enable millions of people to be able to afford to travel to space.

To achieve these goals will require improvements in the following areas:

Expanding Awareness – Expansion of human presence will naturally require expanded monitoring of objects in relevant orbits. No sensors exist to monitor the enormous volume of space between Earth and the Moon (1,000 times the volume to GEO[290]). There are insufficient sensors to provide dedicated

---

[287] Foust, J. (2019). NASA releases ISS commercialization plan. SpaceNews.
[288] Roulette, J. (2021). Axiom names first private crew paying $55 million for a trip to the ISS. The Verge.
[289] Bates, T. (2021). Cislunar Mission Concepts for the Department of Defense. The Space Force Journal.
[290] Holzinger, M., Chow, C. Garretson, P. (2021). A Primer on Cislunar Space. AFRL.

monitoring of rendezvous operations where two vehicles approach each other intentionally, the most hazardous of space maneuvers.

Expanding Information Capacity – NASA currently relies on its Deep Space Network for communication for NASA and partner missions well beyond the Moon's distance from Earth. The capacity of this system is very low, the system is fully tasked with current missions, and is not available for commercial use except in extremely limited cases. Various commercial initiatives are underway for communication throughout the Cislunar volume but NASA employs them only on a limited basis[291] and DOD rarely, and neither has committed to act as a long-term lighthouse customer.

Expanding Autonomy – Logistics missions that support human activity cannot affordably or safely rely on continuous control and monitoring from Earth. Autonomous systems that can be validated for safety will be required to expand to the coming level of in-space operations, both human and robotic.

Expanding Energy Supplies – Both the human missions themselves and the logistics missions that support them will require large quantities of propellant and highly efficient propulsion systems to achieve affordable routine missions. Systems such as high-power electric propulsion, nuclear thermal and nuclear electric propulsion,[292] and Lunar-sourced propellant[293] must all be developed to sustain the spaceflight ecosystem.

KEY INFLECTION POINTS

- **Demonstration of technical and commercial feasibility of personal human transport to near space** by companies such as Virgin Galactic, Blue Origin and SpaceX, and SpaceX.

- **Demonstration of the transport and logistics for human return to the Moon** and establishing a permanent Lunar outpost, base or other sustained presence.

- **Success of routine U.S. commercial, human, space-transport in Cislunar space.**

- **Demonstration of a sustainable market for personal space travel.**

- **Human presence in space passes various thresholds (20, 100, 500, 2000 humans in space simultaneously).**

- **Demonstration of a technically- and economically-feasible commercial human habitat in space** as a destination for personal space travel.

- **Decrease in launch costs below $100/kilogram** for logistics cargoes (not human spaceflight); evidence of a Moore's Law for launch cost.

---

[291] NASA has made use of some commercial networks for Lunar missions (e.g., LRO, LADEE) and certain heliophysics and astronomy missions

[292] Gilbert, A. (2021). Enhancing Military and Commercial Spacepower through Nuclear. The Space Force Journal.

[293] Jehle, A. & Sowers, G. (2021). Orbital Sustainment and Space Mobility Logistics. The Space Force Journal.

- Demonstration of the benefits of and need for human presence in space for manufacturing or resource extraction.
- Demonstration of propellant production on the Lunar surface.

KEY ACTIONS & RECOMMENDATIONS

SHORT-TERM PAYOFF

Sustain commitment to return human presence to the Moon by 2024 with an approach that contributes to advancing beyond exploration toward long-term human presence on the Moon. (OPRs: EOP, Congress, NSpC, NASA, DOC)

Establish as a central national goal establishing a commercially viable, self-sustaining human presence in space and a clear roadmap to achieve this end. (OPRs: EOP, NSpC)

Reduce government competition with commercial SDA capabilities and transition space traffic management systems to using commercial data when available. (OPRs: DOD, DOC)

Develop a joint public-private roadmap for communications within the Cislunar volume that supports human and robotic missions both in orbit and on the Lunar surface. (OPRs: NSpC, DOD, NASA)

Commit to develop a national logistics infrastructure that supports the needs of human missions in Earth and Lunar orbit and on the Lunar surface, as well as propellant delivery from both Earth and the Lunar surface. (OPR: NSpC, EOP)

MID-TERM PAYOFF

Commit to design and build a small-scale, in-space demonstration of a rotating habitat within 5 years initially for exploration but with extension for commercial use. (OPRs: NASA)

DOD specifies the role of human presence in space as a tool for military operations, and DOD determines their role in the defense of U.S. human occupied or operated exploration and commercial systems. (OPRs: USSF)

Deploy the first components of the national space logistics system within 5 years and demonstrate the ability to support human missions in disparate orbits. (OPRs: NSpC)

Space traffic management systems rely on commercially produced data for 50% of their requirements. (OPRs: DOC, USSF)

LONG-TERM PAYOFF

Demonstrate construction and operation of large, life supporting space structures produced and assembled in space from in-space resources. (OPRs: NSpC, DARPA, NASA)

Demonstrate a sustainable human presence on the Moon that directly supports a propellant production facility. (OPRs: NSpC, NASA)

Figure 61: Space Infrastructure Dexterous Robot (SPIDER), a technology demonstration planned for NASA's On-orbit Servicing, Assembly, and Manufacturing 1 (OSAM-1) mission.(Credit: Maxar Technologies)

PERVASIVE SPACE TECH & SUPPLY CHAIN ENABLERS BEYOND LOW EARTH ORBIT

> *"If you build it, they will come."*
> *– FIELD OF DREAMS, 1989*[294]

BACKGROUND

While the majority of commercial, civil, and military spacecraft operate in orbit around the Earth (e.g. LEO, GEO), the volume of space outside of Earth orbit is becoming economically and militarily significant. The expanded orbital regimes that this Cislunar space encompasses offers new opportunities for commercial development, and new responsibilities for military operations. As with other domains, the nation(s) with a superior capability for communication, movement, and sustainment within that domain will dominate its future development.

> *"When established in December 2019, USSF was tasked with defending and protecting U.S. interests in space. Until now, the limits of that mission have been in near Earth, out to approximately geostationary range (22,236 miles). With new U.S. public and private sector operations extending into Cislunar space, the reach of USSF's sphere of interest will extend to 272,000 miles and beyond - more than a tenfold increase in range and 1,000-fold expansion in service volume. USSF now has an even greater surveillance task for space domain awareness (SDA) in that region."*
>
> – USSF-NASA MOU, 21 SEP 2020

Cislunar Space is Big – If we use the GEO radius as a starting scale (42,164 km = 1 GEO), Cislunar distances range out to more than 13 GEO, with Cislunar volume approximately 2,000x larger than the volume inside GEO.

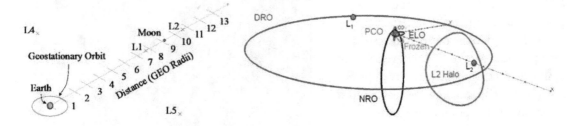

Figure 62: (*left*) Distance model of Earth-Moon system using GEO radii as a basis for scale.[295] (*right*) A sampling of Cislunar orbits including Distant Retrograde (DRO) and Earth-Moon L2 Halo.[296]

---

[294] This is how the public has interpreted and remembered the original movie quote, replacing 'he' with 'they.'
[295] Reproduced from Holzinger et al (2021). A Primer on Cislunar Space. AFRL.
[296] R. Whitley and R. Martinez (2016). Options for staging orbits in Cislunar space. IEEE Aerospace Conference, 2016, pp. 1-9, doi: 10.1109/AERO.2016.7500635.

Cislunar Orbits are Complicated – The two-body (Keplerian) assumption that works quite well in describing orbits of satellites below GEO does not translate well to the Earth-Moon system. Here, three-body effects can cause large deviations from traditional orbit elements (e.g. eccentricity, inclination, etc.) and are not well described by the Two-Line Element (TLE). The result is that trajectories in Cislunar space do not follow easily predictable paths and may include out-of-plane motion and non-circular, aperiodic behavior that is difficult to succinctly describe and visualize (see Figure 62). The diverse landscape of trajectories and orbits in Cislunar space offers an opportunity for spacecraft to take advantage of Cislunar 'high-ground' positions. For example, the Earth-Moon geometry of the L2 Halo orbit provides for Lunar far side operations and/or communication relays.[297]

CURRENT STATE

The past year has been one of rebuilding after significant timeline disruptions for Cislunar projects due to COVID-related impacts and selected parts shortages. The global supply chain for space-related components has struggled to recover at a rate fast enough to keep up with resurgent demand. As project designers continue to emphasize flight heritage to mitigate technical risk, there is a lack of space-proven components in certain segments of the global supply chain, and significant long-lead delivery times in others. This is likely to have adverse flow-down impacts to program schedules and milestones.

"My area of operations includes the Lagrange points, strategically vital way stations of sorts where gravitational forces of the Earth and the Moon balance each other out. These gravity wells are ideal for positioning spacecraft, where they can remain indefinitely with only using a small amount of fuel. A maritime analog to the Lagrange points would be the strategic importance of several very small islands of the Pacific."

– GEN JIM DICKINSON, USSPACECOM Commander, 3 August, 2021[298]

China Continues to Expand and Develop its Presence in Cislunar Space -- Since the publishing of SSIB'20 report, China completed the first sample return from the Moon in 40 years with the Chang'e 5 lander. Meanwhile, the new U.S. administration has maintained the commitments to the Artemis program, which leverages significant commercial capability to deliver humans to the Lunar surface and achieve several scientific goals. With the global competitive environment dominated by Chinese and U.S. Government civil Lunar ambitions, the private sector remains somewhat tied to these efforts. Investors remain skeptical of commercial ventures in Cislunar space, with the majority of private investment in space going towards platform technologies (e.g. propulsion), launcher development and launch services, satellite development and ground system infrastructure and a variety of activities in low Earth orbit.[299]

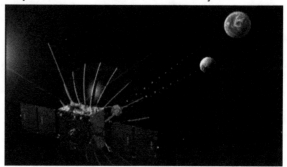

Figure 63: China's Queqiao or "Magpie Bridge" communications relay satellite in L2 (Credit: CNSA)

---

[297] Farquar, R.W. (1971). The Utilization Of Halo Orbit In Advanced Lunar Operations. NASA.
[298] Hitchens, T. (2021). USSPACECOM Head Touts Space, High Seas Parallels. Breaking Defense.
[299] Space Capital (2021). Space Investment Quarterly Q2 2021. Space Capital.

Cislunar Starting Points – Communication relays and Position-Navigation-Timing (PNT) infrastructure have been identified as key enabling services for economic development of Cislunar space. The need for reliable Cislunar space domain awareness is also a key to developing a vibrant Cislunar economic sphere. A small number of startups are active in this area, but most VC firms are first waiting for the U.S. Government to commit to a strategy for Lunar infrastructure.

Mobility Technology – A small number of businesses are capable (or nearly capable) of ex-GEO operations. Spacecraft that operate in Cislunar space are typically one-off designs, ideally with most subsystems having proven space heritage. Even with advanced chemical and electric propulsion systems, delta-V capability is still a limiting factor for several operational concepts (e.g. TLI tug, Lunar cycler, on-orbit servicing, etc.). Both military and civil agencies are committed to development of nuclear thermal propulsion (e.g. DRACO, PELE), which promises nearly double the performance of chemical propulsion systems. Additionally, private investment in buses/tugs/cycler technology development is active, as investors recognize the added market value of advanced propulsion in both Earth orbit and Cislunar space.

Figure 64: NASA concept illustration for a Mars transit habitat and nuclear propulsion system
(Credit: NASA)

Growth of New Markets – On-orbit servicing, assembly, and manufacturing are not pervasive technologies in Earth orbit and are largely one-off technology-demonstration projects. These efforts are still mostly government dominated (e.g. RSGS), although some small businesses are gaining momentum in this area, with Orbit Fab launching the first commercial on-orbit fuel depot demonstrator in 2021.

KEY ISSUES & CHALLENGES

The effects of the COVID-19 pandemic continue to impact space-related supply chains. The supplier base is not ready for rapid expansion and growth into Cislunar space. If government and private investors rapidly increased funding into the space industrial base for Cislunar development, it would take several years before the supplier base could respond to that increase in demand. This is primarily due to limits on how quickly the space industry can ramp up highly-specialized hiring, facilities, equipment, design and integration capabilities. New technologies, especially those related to space, require sustained investment, policy updates, and focus over years to be successful.

With the proliferation of launch providers, getting something into orbit is no longer the bottleneck in space development. Satellite integration and testing facilities and regulatory/licensing approvals are becoming the limiting factor in rapid test and evaluation iteration of space technology.

80% of all space launches in 2020 were commercial. However, U.S. companies earned only 40% of the $5.3 billion in global revenue.

- BRYCE TECH[300]

Small Business Barriers – Participation in U.S. Government-sponsored innovative technology development (NASA and/or DOD) can place a burden on small businesses to field adequate proposals, deal with extensive contracting processes and reporting requirements, maintain cyber security compliance and in some cases obtain personnel security and facility clearances. Feedback from industry indicates that the many innovative technologies related to Cislunar space are also dual-use or ITAR/EAR-restricted, which discourages small business from participating.

A Lack of a Defined Market -- A lack of a defined market for Cislunar space infrastructure makes it challenging for commercial companies looking for private investment in this market. A lack of interface standards also hampers the proliferation of in-space manufacturing and servicing capabilities. Innovative technologies like nuclear propulsion face regulatory and licensing challenges and a lack of testing facilities. Over-reliance on continued government support as the customer or tech-development lead challenges the viability of an independent commercial market for on-orbit manufacturing.

Risk Posture – Both government and commercial Cislunar space stakeholders have developed an aversion to technical risk from decades of building exquisite, one-off spacecraft. For government-led programs, this has created an over-reliance on low-risk, high flight-heritage components and a stagnation of spacecraft capabilities. On the purely commercial side, insurance providers are unwilling to take on technical risk, disincentivizing companies to develop and launch high-risk, high-reward technologies. The public-private partnership model (e.g. NASA's Commercial Lunar Payload Services (CLPS) program) is bridging this gap by melding top-level government mission requirements and funding with emergent commercial mission capabilities.

Figure 65: Capture and mining of asteroids ranging from 10 to 40 meters in diameter (Credit: TransAstra)

Despite these challenges, the space industrial base is in a good position for a slow growth profile. Unfortunately, slow growth into Cislunar space may not be fast enough to maintain U.S. dominance in this domain.

---

[300] Bryce (2021). State of the Satellite Industry Report. Bryce Tech. Retrieved from https://www.brycetech.com

KEY INFLECTION POINTS

> *"In 2001, China publicly announced its specific plans to send missions to the Moon under the Chang'e program. This ambitious program included developing communications infrastructure, robotic missions to the far side of the Moon, sample returns, and eventually human missions. These missions have implicit and explicit dual-use military applications. Early U.S. attention to this plan and its implications would have driven us to realize the need for (at a minimum) a Cislunar Space Domain Awareness (SDA) mission. A small investment a decade ago would have led to mature capabilities today."*
> — USSF SPACE FUTURES WORKSHOP REPORT, 2021

In considering how Cislunar might evolve,[301] several futures are worth contemplating which vary according to the strength or weakness of U.S. leadership, and the posture taken by the U.S. Government toward industry.

NASA-dominated - 80% of cash flowing into Cislunar economy comes from/through NASA	**Lunar hubs** - 3 to 5 locations on-orbit and on surface are of primary interest
Stare-down / rugby play - Two opposing countries or teams compete for Cislunar dominance	**Lunar wildcatting** - High value locations on-orbit and on surface are identified; competition to locate and secure these sites drives Cislunar activity, especially commercial
Cislunar space ceded - U.S. Cislunar support wanes; no longer a domain of interest	**Nothing** - Support for Cislunar activities collapses worldwide

Table 4: Possible Cislunar Futures.[302]

- **Communications and Data** - First commercial ex-GEO comm relay and government commitment to use capability; Government commitments to Cislunar SDA infrastructure
- **Mobility** - First ex-GEO demonstration of refueling; Demonstration of advanced propulsion capability would be a key enabler for quick expansion and proliferation of Cislunar infrastructure
- **Cislunar Servicing and Assembly** - First commercial Business-to-Business spacecraft servicing in Cislunar space would be key indicator that commercial market is feasible without reliance on government contracts
- **Cislunar Manufacturing** First launch of commercial manufacturing capability (non-ISS); first commercial sample return to Earth of Cislunar-manufactured item – both are indicators of market viability and could spur further commercial investment and development
- **Lunar Outpost** construction of Lunar outposts by any nation or coalition. Degree of ex-GEO Cislunar activity ultimately depends on the level of Lunar surface activity, including development of permanent outposts, ISRU facilities, mining activities, scientific, or astronomical facilities.

---

[301] Duffy, L. & Lake, J. (2021). Cislunar Spacepower, The New Frontier. The Space Force Journal.
[302] SSIB'21 Pervasive Space Tech & Supply Chain Enablers Beyond Low Earth Orbit outbrief presentation.

KEY ACTIONS & RECOMMENDATIONS

SHORT-TERM PAYOFF

Continue procurement of commercial products and services in Earth orbit and Cislunar space. This not only provides short term benefits to end-users, but it also helps to de-risk business cases and encourages further commercial investments in space. (OPRs: DOD, NASA)

Commit to Cislunar infrastructure by accelerating adoption of interface and communication standards, providing PNT services or incentivizing commercial PNT and communication services, establishing a Cislunar domain awareness capability, and providing data security standards. This would remove some of the uncertainty in the commercial market and support rapid industry response. (OPRs: EOP, NSpC)

"The universe is an ocean, the moon is the Diaoyu Islands [Senkaku Islands, East China Sea], Mars is Huangyan Island [Scarborough Shoal, South China Sea]. If we don't go there now even though we're capable of doing so, then we will be blamed by our descendants. If others go there, then they will take over, and you won't be able to go even if you want to. This is reason enough."

— YE PEIJIAN, Head of China's Lunar Exploration Program, 2018[303]

MID-TERM PAYOFF

Government buys Lunar internet/data service which will provide incentive for commercial companies to develop data infrastructure. Clarity of government intentions will help industry refine and apply their strategies. (OPRs: DOD, EOP, NSpC)

Support advanced propulsion RDT&E offers the ability to place and maintain infrastructure assets, move materials between orbits, enable on-orbit manufacturing and satellite servicing; mobility gained by increased delta-V adds to feasibility of new mission sets in Cislunar space. (OPRs: DOD, USSF, NASA, DOE)

Use of prizes for DOD-relevant technologies and goals; major awards for accomplishing key milestones would shift focus from 'requirements'-driven acquisition towards 'mission'-driven industry innovation. (OPRs: DOD, DOC, NSpC)

LONG-TERM PAYOFF

Remove barriers for small business by incentivizing primes to assist small businesses with cybersecurity compliance, ITAR/EAR, security issues, etc. to encourage more participation from industry in dual-use technology-development efforts. (OPRs: DOC, DOD, FAA)

Re-examine risk stance to encourage innovation with government commitment to fund transition of low-TRL, high-risk, high-reward technologies with potential applications in Cislunar space. The Government may also provide loan guarantees or tax incentives to businesses that are developing innovative/agile technologies for space. (OPRs: DARPA, DOD, NASA)

---

[303] Hong, P. (2018). China's Looming Land Grab in Outer Space. The Daily Beast

EPILOGUE

"The U.S. in general, and the Space Force specifically, face great challenges anticipating, shaping, and preparing for the future of space to support U.S. national power."

— USSF SPACE FUTURES WORKSHOP, 2021

CONCLUDING THOUGHTS

Space Industrialization is Here - Space industrialization is not some far-off dream. A vibrant space economy is extant. Existing companies are advancing the technology for new global markets, for in-space manufacturing and resource extraction, and for supporting logistics, transportation and communication and navigation systems. America, or its rivals, will be setting precedents that will condition the entire system. Urgent attention is required to set the conditions for the future of the domain.

Figure 66: Harvesting of water ice from Lunar cold-traps on the Moon (Credit: James Vaughan).

Space Offers a Golden Opportunity - The space domain is evolving rapidly, and today space is ripe with opportunity to establish first-mover advantage. It is among the areas most sensitive to initiative in strategic competition. Now is the time to press our advantages in commercial space to reap enduring economic and soft power advantages.

A Robust U.S. Industrial Base is a National Imperative - Participants share a common theory of 'causing national security and building enduring military advantage': *The secret sauce of America's strategic power has been the conscious and deliberate nurturance of strategic industries for seafaring and aviation.* Strategic strength comes from economic and industrial strength, enabled by the transportation modes, which give access to a strategic domain and manufacturing capability to field, at-scale, power military platforms when required. Recognizing that access and manufacturing is the source of strategic strength, *government mobilizes and subsidizes the creation of dual-use industries.* Therefore, neither the national leadership nor the Department of Defense can look at the growth of the space industry as an externality. Using what Albert Einstein called the 'magic of compound interest,' the DOD can make a small one-time investment to catalyze a self-sustaining and scalable industry, which it can draw upon sometime in the uncertain future. This requires a change in mindset within the DOD.

What is to be done immediately? - A great deal of actionable items have been proposed by the workshop participants which would allow the United States to align its vibrant space sector with its larger policy initiatives. *This is the year to tie in climate change, power, resources, on-orbit assembly and manufacturing into our broader grant strategy to prevail in strategic competition.*

Working Ahead for Next Year - If policy is to move in the directions indicated in this report, much of the work for next year's State of the Space Industrial Base is suggested. Participants from industry, government, and academia need to arrive with more specific proposals for climate change, power, space resources and on-orbit assembly to make policy immediately actionable.

APPENDIX A
WORKSHOP PARTICIPANTS

LIVE PARTICIPANTS

Steve Altemus, Intuitive Machines
Michael Alvarez, Ecliptic Enterprises Corp.
Julie Baker, Ursa Space Systems Inc.
Payam Banazadeh, Capella Space
Dave Barnhart, Arkisys, Inc.
Klay Bendle, Defense Innovation Unit
Scott Bennett, PredaSAR Corporation
Nathan Bishop, Nanorakcs
Stephanie Blake, Orbital Effects
Severin Blenkush II, Space RCO
Michael Bloxton, Nebula Space Enterprise, Inc
Heinrich Bornhorst, BlackSky
Denise Brickley, RESPEC Company, LLC
Austin Briggs, Inversion Space
Daniel Brophy, Capella Space
Craig Brower, Orbital Insight
Brig. Gen. Steven Butow, Defense Innovation Unit
Kristin Burke, ODNI
Marta Calvo, HelicitySpace Corp.
Daniel Ceperley, LeoLabs
Dan Chi, USAF OCEA
Dr. Thomas Cooley, AFRL/RV
Paul Damphousse, Calspan Holdings
Jill Davis, AFRL/RV
Arial DeHerrera, New Space New Mexico
Casey DeRaad, New Space New Mexico
Derek Doyle, AFRL Space Vehicles Directorate
Laura Durr, USSF
Bailey Edwards, SpinLaunch, Inc.
Robert Eidsmoe, Space RCO
Daniel Faber, Orbit Fab
Cain Fallin, Northrop Grumman
Navid Fatemi, SolAero Technologies
Col. Eric Felt, AFRL/RV
Ethan Feuer, The Aerospace Corporation
Justin Fiaschetti, Inversion Space
Dave Fischer, Astroscale U.S. Inc.
Brien Flewelling, ExoAnalytic Solutions Inc.
Richard French, Rocket Lab
Renee Frohnert, L3Harris Technologies
Rob Gabbert, Gabbert Consulting Co. LLC
Victor Gami, Tau Technologies LLC
Peter Garretson, Apogee Engineering LLC
Edwin Geisel-Zamora, X-Bow Launch Systems
Nate Glover, Northrop Grumman Tactical Space Systems
Javier Gomez, RSC-NCR
Michael Good, Lockheed Martin
Dr. Kelly Hammett, AFRL/RD
Dr. David Hardy, Apogee Engineering LLC
William Hargus, AFRL/RQ
Robbie Harris, Nanoracks
Adam Harris, Orbit Fab
Erika Hecht, New Space New Mexico
Dan Hegel, Blue Canyon Technologies
Carrie Hernandez, Rebel Space Technologies, Inc.
Michael Hodge, Lynk Global
Lars Hoffman, Rocket Lab, USA
Scott Jacobs, Astranis Space Technologies Corp.
Theodore Johnston, Boundary Stone Partners
Dennis Kater, Quantum Research International
Fred Kennedy III, Momentus
Dr. Barry Alan Kirkendall, Defense Innovation Unit
Patricia Knighten, Arrowhead Center, NMSU
Katherine Koleski, Defense Innovation Unit
Justin Koo, AFRL/RQRS

Andrew Kwas, Northrop Grumman Space Systems
Matthew Lamanna, Deloitte
Cameo Lance, Rhea Space Activity, LLC
Kendra Lang, Verus Research
Eric Lasker, Varda Space Industries, Inc.
Juli Lawless, Redwire Space
Dan Lopez, Arkisys, Inc.
Erik Luther, CesiumAstro, Inc.
Scott Maethner, New Space New Mexico
Jason Marino, OUSD R&E
David Marsh, Nanoracks
John Mayberry, The Aerospace Corp
Tom McCarthy, Motiv Space Systems
AJ Metcalf, AFRL/RV
Kyla Miller, AFRL/RX
Scott Milster, AFRL/RV
John Mobery, LeoLabs
Lt. Gov. Howie Morales, New Mexico Lt. Governor
Nate Moser, Quantum Research International
Kent Nickle, Axient LLC
Steve Nixon, SmallSat Alliance
Brig. Gen. John Olson, USSF
Greg Oslan, Self
Matthew Palm, Microsoft Federal
Devon Papandrew, STOKE Space
Dennis Poulos, Defense Innovation Unit
Chris Power, Hadrian
Chris Quilty, Quality Analytics
Atif Qureshi, Maxar Space Robotics
Dennis Ray Wingo, Skycorp Incorporated
Bill Raynor, U.S. Naval Research Laboratory
Benjamin Reed, Quantum Space
John Reeves, Viasat
CDR Jeremy Reynard, Defense Innovation Unit
Jeff Rich, Xplore Inc.
Jared Rickewald, SpaceLogistics LLC, Northrop Grumman

Rex Ridenoure, Defense Innovation Unit
Dr. Gordon Roesler, Defense Innovation Unit
Brian Rogers, Rocket Lab USA
Capt. Lauren Rogers, Defense Innovation Unit
Alan Rosenthal, The Aerospace Corporation
Sean Ross, AFRL/RD
Maj. David Ryan, Defense Innovation Unit
Shey Sabripour, CesiumAstro, Inc.
Erik Sallee, Intuitive Machines
Tony Samp, DLA Piper
Gregory Sanford, Redwire
Venke Sankaran, AFRL/RQ
John Santacroce, Northrop Grumman
Maj. Gen. Jay Santee, Strategic Space Operations, Defense Systems Group, Aerospace Corp.
Adrienne Schaab, AFRL/RXME
Adam Schillfarth, USNC-Tech
Greg Seeley, Guardian Consulting
Brandon Seifert, USNC-Tech
Dr. Joel Sercel, TransAstra Corporation
Dr. Rogan Shimmin, Defense Innovation Unit
Keith Shrock, BlueHalo
Maj. Russell Stanton, Defense Innovation Unit
Karl Stolleis, AFRL/RV
Chris Thayer, Motiv Space Systems
Eric Truitt, PredaSAR Corporation
Rick Tumlinson, Space Fund/Space Frontier Foundation/Earth Light Foundation
Ben Urioste, AFRL/RV
Mandy Vaughn, GXO
Paolo Venneri, USNC-Tech
Annijke Wade, Cognitive Space
Dr. Ryan Weed, Defense Innovation Unit
Pete Wegner, BLACKSKY
Stacie Williams, AFOSR

VIRTUAL PARTICIPANTS

Jack Ackerman, USNC-Tech
Austin Albin, Velos
Wilmer Alvarado, Khronos Corporation
Eric Anderson, And One Technologies
Paul Armijo, Avalanche Technology
Jason Aspiotis, Axiom Space Inc.
Richard Aubert, Defense Contract Management Agency
Josh Baty, LytEn
Kari Bingen, HawkEye 360
Dr. Bruce Cahan, Stanford/Urban Logic
Elliot Carol, Lunar Resources, Inc.
Laurie Chappell, MDA Systems Inc.
Brad Clevenger, SolAero Technologies Corp
Judith Connelly, USCG R&D
William Cook, Missile Defense Agency
Keith Costa, CNA
Guy de Carufel, Cognitive Space
Maj. Joseph Dechert, USAF - AFRL/RQ
Christopher DeMay, TrustPoint
Jessica Downey, Precision ISR
Sean Duggan, Office of Sen. Martin Heinrich
Molly Fischer, Stellar Science
Dennis Fitzgerald, CMTC
Charles Fletcher, SIBWG NASA
Alex Friedman, AFRL/RV
Karen Gaffin, USAF Office of Commercial & Economic Analysis
Capt. Nathan Gapp, Defense Innovation Unit
LTC Alexander Garcias, Defense Innovation Unit
Dr. Michele Gaudreault, USAFA
Dr. Bill Goodman, Goodman Technologies LLC
Johnnie Hobbs, Parsons support contractor for Missile Defense Agency
Luca Ispirescu, IDEAS-TEK
Peter Jackson, OUSD(A&S)
Greg Karahalis, Missile Defense Agency

Bernard Kelm, U.S. Naval Research Laboratory
Jim Keravala, OffWorld
Patrick Kong, USAF OCEA
Benjamin Kron, York Space Systems
Kali Kuthra, Velos
Michael Laine, LiftPort Group
Dr. Bhavya Lal, NASA

Robert Lasky, AstronetX
Gerard Lebar, Northrop Grumman
Tom Loftus, Razor's Edge Ventures
Autumn Lorenz, NGB Space Operations Directorate
Jon Lutz, Honeybee Robotics
John Lymer, MDA
Mike Manor, Sierra Nevada Corporation
Mark Massa, Atlantic Council
Todd Master, Umbra
Jessica McBroom, National Security Council

Kevin McClellan, Apollo Fusion Inc.
Dan McCullough, Lockheed Martin Space, Military Space
Michael Mealling, Starbridge Venture Capital
Steven Meier, NCST Code 8000
Hugh Mills, Innoflight
Michael Moran, PredaSAR Corporation
Terry Mosbaugh, Emergent Space Technologies
Christopher Murply, U.S. Navy
Alan Perkins, Missile Defense Agency - Industrial and Manufacturing Technology
Dr. Shawn Phillips, AFRL Rocket Lab
Elfego Pinon, Emergent Space Technologies, Inc.
Bruce Pittman, Skycorp Inc.
Barry Ressler, International Space Medicine Consortium Inc.
Lisa Rich, Xplore Inc.
Dr. Mir Sadat, Space Force Journal
Jim Schwenke, SpaceLink
Stan Shull, Alliance Velocity, LLC
Julia Siegel, Atlantic Council

Scott Singer, Missile Defense Agency
Pavnette Singh, Defense Innovation Unit
Nicholas Sramek, The Aerospace Corp.
Clementine Starling, Atlantic Council
Gregory Stottlemyer, Advanced Product Transitions Corp
Debi Tomek, NASA
Kathy Trimble, U.S. Air Force
Andrew Tucker, Guardian Space Technology Solutions
Eddie Tunstel, Motiv Space Systems, Inc.
Erin Vaughan, AFRL/RV
Christopher Werth, Aerospace Corporation
Bill Woolf, Space Force Association
George Xiao, Microcvd Corporation

APPENDIX B
PREVIOUS REPORTS & KEY RECOMMENDATIONS

Space Power Competition in 2060: Challenges and Opportunities
Report on the Space Futures Workshop 1A

9 Mar 2019

Distribution D:
Authorized to the Department of Defense and U.S. DOD contractors only

- A long-term, national space strategy integrating civil, commercial and national security space lines of effort must be developed to retain the U.S.' dominant and leadership position in the emerging future of space. This strategy must account for the possible space futures developed in the workshop.
- The overall strategy must address how the national security establishment will defend the full range of national interests in space—not just the services that are provided directly for national security.
- AFSPC must commit the resources to continue to lead in completing the remaining steps in the process to define these futures, as key inputs to the strategy, and to determine their implication for present and future defense strategy. U.S. Space Command should similarly commit resources to this end as part of their strategic and operational execution missions.
- Essential capabilities and technologies to enable positive future outcomes must be developed by the whole of government. An investment, policy, and regulatory strategy must be pursued to ensure those capabilities.
- To maintain our technological advantage in space, the nation must commit to continued investment in science and technology related to the rapidly changing global space environment.

Download

State of the Space Industrial Base: Threats, Challenges and Actions
A Workshop to Address Challenges and Threats to the U.S. Space Industrial Base and Space Dominance

30 May 2019

Distribution A:
Approved for Public Release. Distribution Unlimited.

- Upgrade of our own methodologies such as shared, trusted supply chains and interoperable technology standards that accelerate viable commercialization of the space economy.
- Develop a more flexible, U.S.-led markets for space capabilities that spread the risk, increase the pool of investors and establish U.S. leadership in setting a framework for a Cislunar commercial space economy that creates wealth and security with our allies and partners who share our common norms and values.
- Changes in U.S. Government procurement and licensing processes and other regulations to eliminate unnecessary delays and micro-management of the space industrial base's ability to rapidly deliver next generation space capabilities, and to enable early U.S. investment in capabilities from emerging, innovative elements of the space industrial base.

The Future of Space 2060 & Implications for U.S. Strategy
Report on the Space Futures Workshop

5 Sep 2019

Distribution A:
Approved for Public Release. Distribution Unlimited.

Download

- The 2060 space world will be highly complex and diverse as to the number of state and non-state actors, their capabilities, and their interests.
- Commercial space presents unique issues as to ownership and sovereignty that, if not resolved, could lead to commercial space entities as independent or semi-independent space powers, resulting in significant opportunities and challenges to U.S. space power.
- Space power will be widely distributed, making it impossible for any one nation or entity to have predominant space power in the civil, commercial, and military domains.
- The diversity and distribution of space power enables a wide range of alliances, partnerships, and shared interest. These relationships will be diverse and vary with time as the interest and capabilities of space faring entities develop and change. This complexity poses significant challenges to the U.S. to maintain and exercise the space power needed to protect its interests in space and in the terrestrial sphere.

State of the Space Industrial Base 2020
A Time for Action to Sustain U.S. Economic & Military Leadership in Space

July 2020

Distribution A:
Approved for Public Release. Distribution Unlimited.

Download

- The world stands at the threshold of a new era in space. Other nations are challenging the U.S. for leadership of this next space age. Success in this long-term strategic competition requires that the U.S. seamlessly integrate multiple elements of national power.
- This report clearly delineates the state of the space industrial base, its challenges and issues, and provides six overarching recommendations to U.S. policymakers and four to space industry leaders.
- It further provides detailed analysis of sub-areas of the space industrial base and areas affecting the base, specifying 39 actions as to what must be done, who must do them, and the timeframe in which they must be accomplished.
- The most important actions for government and industry include: the promulgation of a whole-of-government, "North Star" top-level vision and strategy for space industrial development; the DOD develop plans to protect, support, and leverage commerce in space; economically stimulate the industry by executing $1 billion of existing DOD and NASA funding through space bonds and a Space Commodities Exchange; create wealth and security with allies and partners that share our common norms and values; create and fill more than 10,000 Science Technology Engineering and Math (STEM) jobs domestically; and the USSF works closely with space industry entrepreneurs and innovators to

develop government-commercial technology partnerships that support U.S. commerce and national security in space.

Defining the Road to 2035-45 USSF Capabilities
Report on the USSF Space Futures Workshop 2a

5 Aug 2021

Distribution D:
Authorized to the Department of Defense and U.S. DOD contractors only

- The U.S. must develop and execute a grand strategy for space recognizing space's importance and enhancing our advantages. This strategy must encompass near-term terrestrial-focused power and a long-term focus on Cislunar expansion and beyond as a domain in itself for human action.
- A robust/competitive U.S. Space Industrial Base is essential to maintaining the U.S. as a preeminent space power, but its competitive advantage is threatened by increasing globalization of space industrial capabilities.
- By 2040, USSF missions may include: increased space information services; projection of offensive and defensive operations in space and from space to other domains; dynamic offensive/defensive operations and transport across the Cislunar domain to ensure freedom of civil, commercial, military operations; environmental monitoring, stewardship and debris clean-up; protection of critical space national infrastructure; enforcing space law & norms of behavior; Search & Rescue / Personnel Recovery (PR) / Non-Combatant Evacuation (NEO); and planetary defense.
- Space Science and Technology most affecting the future of space include: launch system reusability/diversity for space access to decrease cost/increase flexibility; bus & payloads to lower development deployment costs of space internet & communications; bus & payloads to lower cost and increase diversity of sensing modes from space, to both space and terrestrial domains; bus, robotics and AI/ML/Autonomy advances for on-orbit servicing, assembly and repair; AI/ML/Autonomy for offensive/defensive space operations; in-space propulsion and navigation for increased on-orbit range, speed and maneuverability; launch and transport systems for human access and mobility in space; cyberspace protection of space capabilities; economically viable space manufacturing; and efficiency/cost of space-based solar power.

This page was intentionally left blank.

APPENDIX C
2021 DIU SPACE PORTFOLIO COMPANY SURVEY[304]

EXECUTIVE SUMMARY

BY THE NUMBERS:

Total respondents representing the Space Industrial Base (SIB)	57	100.0%
Companies with 100 or less employees	43	74.4%
Classify as "non-traditional" defense contractors[305]	41	71.9%
More than 90% of revenue from within U.S. only	34	59.6%
Experiencing supply chain issues leading to delays of > 8 weeks	22	38.6%
≤12 months of 'runway' or funding based on current burn rate	23	40.0%

Table C-1: Survey respondent demographics

GENERAL OBSERVATIONS

- Most companies:
    - Cited unpredictable U.S. Government demand as the greatest hurdle to growth.
    - Cited export controls and ITAR-restrictions in their top-4 hurdles to growth.

SURVEY

This survey was initiated on 22 June 2021 as a follow-up to DIU's 2020 COVID-19 Impacts Survey. Participating companies are fairly well funded with familiar names and <100 employees (median value). The survey questions were formulated to assess risk, identify challenges and measure effects on the greater U.S. space supply chain. Approximately 50% of the surveyed companies responded. The 2021 survey covered additional topic areas versus the previous survey, including hiring, working with the U.S. Government, revenue & investment, supply chain impacts and policy.

---

[304] Survey responses reflect the views of the participating companies representing the U.S. space industrial base and do not necessarily reflect the official position of the USSF or Department of Defense.

[305] A company is considered a non-traditional defense contractor if any of the following apply: (1) small business exempt from 41 USC Chapter 15 Cost Accounting Standard (CAS) requirements; (2) exclusively perform contracts under commercial procedures; (3) exclusively perform under firm-fixed-price (FFP) contracts with adequate price competition; and (4) performed less than $50 million in CAS covered efforts during the preceding cost accounting period.

FINDINGS

FINDING 1: Demographics and Personnel

Fifty-seven U.S. commercial space companies were surveyed in June 2021 with respondents spanning 11 states with significant commercial space activity across the United States (See Figure C-1).

Figure C-1: Distribution and density of Survey respondents (Credit: DIU).

The companies surveyed were primarily non-traditional small businesses with most companies having between 26 and 100 employees, with most companies (~75%) having between 1 and 100 employees. This indicates that the Space Industrial Base (SIB) is predominantly small, early stage businesses aligned with demand signals from the government and private industry. Most have tremendous potential for growth in a globally competitive new space market.

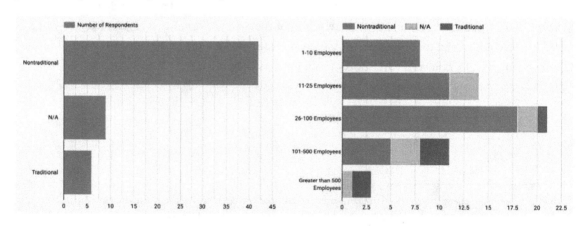

Figure C-2: Demographics of Survey Respondents (Credit: DIU).

Survey respondents were asked to rate on a scale of 0 (declined significantly) to 10 (increased significantly) how their ability to hire has changed over the course of the COVID-19 pandemic. For the most part, hiring difficulty has not changed significantly over the course of the pandemic. However, interesting trends emerge when analyzing responses on a per-state basis. We can see that companies based out of New Mexico and Florida have seen a marked increase in difficulty for hiring new employees, whereas states such as Colorado and New York have found it much easier to recruit employees. This aligns with larger trends in the technology sector as well as overall COVID-19 U.S. population emigration.[306]

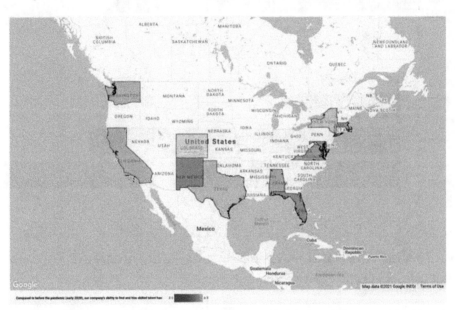

Figure C-3: Reported ability of space companies to hire by state (Credit: DIU).

FINDING 2: Companies are better positioned financially in 2021 versus 2020

The uncertainty of the global pandemic weighed down many companies in early 2020. Venture capital and the public markets have responded positively by investing more dollars into commercial space companies over the past year than in any year prior.[307] It is worth noting, however, that a majority of today's most successful commercial space startups have government business in the form of prototype, procurement and service contracts. Investors took note of this fact and observed that the non-dilutive government money did not "dry up."[308]

In sharp contrast to last year's survey, most companies (~86%) reported a comfortable amount of financial runway for the financial year. Eight companies reported having less than six months of remaining runway, of which half had to cut back on expenses and personnel, with three others having to raise capital in order to maintain their current rate of growth.

---

[306] Patino, M. et al (2021). More Americans Are Leaving Cities, But Don't Call It an Urban Exodus. Bloomberg.
[307] Space Capital (2021). Space Investment Quarterly Q4 2020. Space Capital
[308] Grush L. (2020). How the space industry is weathering the coronavirus pandemic. Verge.

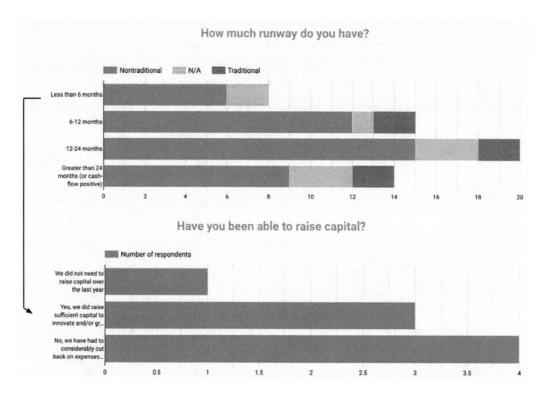

Figure C-4: Reported financial health and private fundraising status (Credit: DIU).

FINDING 3: The Small Business Innovation Research (SBIR) program is a popular way to obtain funding for R&D and capital expenditures

Slightly more than half of the companies surveyed (~56%) have applied for, and successfully received, at least one SBIR grant throughout their company's existence, with an additional 5% planning on applying for one in the next twelve months.

The primary use-case for SBIRs are to fund capital expenditures and technology development. SBIRs were also seen as attractive markers of a government market demand signal to show private investors. However, respondents noted the difficulty of actually applying and getting a SBIR to be high, with many calling the process "difficult", "confusing", and "not user-friendly." Companies noted a "valley of death" beyond Phase 2, due to the difficulty in connecting their technology to a program of record, stalling their traction with the DOD.

FINDING 4: A quarter (26%) of respondents are considering SPACs as a viable exit strategy for their companies

All of these respondents are non-traditional vendors.

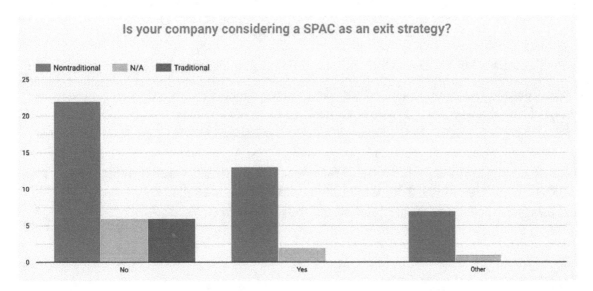

Figure C-5: Reported SPAC exit strategies (Credit: DIU).

The primary cited reason for considering a SPAC is immediate access to capital, along with freedom from recurring fundraising activities.

Figure C-6: Commercial space companies with DOD contracts achieving Unicorn status (Credit: DIU).[309]

---

[309] In business, a 'unicorn' is a privately held startup company valued at over $1 billion. Companies reported here have performed significant prototype work with the DIU Space Portfolio and other DOD partners.

FINDING 5: Two thirds of respondents (65%) are experiencing supply chain delays, with 39% experiencing delays of eight weeks or more

The COVID-19 pandemic created uncertainty in the demand for everything from cars to computer-monitors, gaming consoles, personal electronics, and most things that contain microprocessors. This highlighted the existing tightness in global semiconductor supply by bringing it to a near-critical level globally.[310] On the supply side, semiconductor manufacturing was hampered by COVID-outbreaks and water shortages in Taiwan, where almost half the world's semiconductors are manufactured.

Of the 38 space-industry respondents who said they were experiencing 'downstream-supply issues / delays of critical components', 26 cited electronics or electronic components and 17 of those specified "Microchips" or "FPGAs" as the component at issue. The shock is unlikely to resolve quickly. Supply uncertainty will beget artificial demand spikes as companies seek to build inventory to buffer against future shortages, like shoppers buying more toilet-paper than usual at the sight of empty shelves. Adding capacity takes years and, while necessary, won't come in time to address the current state. Some project significant shortages are likely through the beginning of 2023.[311]

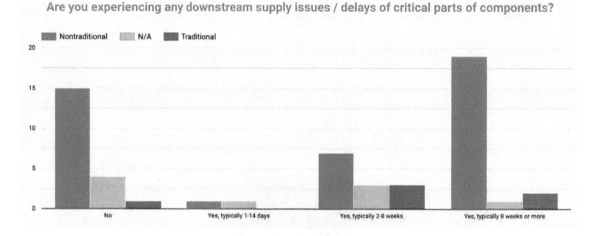

Figure C-7: Downstream supply chain issues and/or delays of critical parts (Credit: DIU).

FINDING 6: There is a renewed interest and demand in radiation-hardened, high reliability, and high performance microelectronics.

Survey respondents cited radiation-hardened electronic components as the second-most in-demand component whose improved availability would positively impact development and production of space-industry product offerings. Specific responses included rad-hardened microprocessors and solar-cells. With the SSIB moving towards commercial off-the-shelf (COTS) hardware in recent years, the demand for rad-hardened electronics signals a potential approaching saturation point for applicability of COTS electronics for all space needs. In-space mobility, logistics and servicing will enable greater use of radiation tolerant (vs hardened) parts in replaceable, modular components.

---

[310] King, I. et al (2021). How a Chip Shortage Snarled Everything From Phones to Cars. Bloomberg.
[311] Supplyframe (2021). Supplyframe Commodity Intelligence Quarterly Points to Increasing Risk and Supply Constraints for Manufacturers. Retrieved from https://supplyframe.com/

FINDING 7: The government's demand signal was deemed as by 42 of the 57 survey respondents (~74%) which was seen as a hurdle to future growth

Specific challenges included access to key points of contact in government agencies and/or the responsiveness of those POCs, as well as the need for a security clearance and/or an existing contract in order to conduct substantial business development activities. Others felt the government needed to make more long-term commitments to support long-term fundraising and investment. Additionally, many vendors expressed that the contracting process remains lengthy and cumbersome. Some recommended leveraging fully-negotiable Other Transaction Agreements, such as DIU's Commercial Solutions Opening (CSO), to get vendors under contract more rapidly and flexibly (i.e. allowing vendors to retain intellectual property).

FINDING 8: Many vendors felt ITAR was a major hurdle to growth

A significant number of respondents cited concerns with the International Traffic in Arms Regulations (ITAR)[312] as it presents a major hurdle to growth in a highly competitive global space market. Specifically, ITAR impedes speed of access to market because of the resources required to interpret its rules, the difficulty getting a clear answer from the government, and the major limitations on vendors' total addressable market. Concretely, vendors felt pressure from competitors in the European Union, with some highlighting competitors who marketed themselves as "ITAR-free." The classification of "Space Servicing" under Part 120.7 of the ITAR regulations was described as "incredibly broad and subject to subjective [sic] interpretation."

---

[312] For more information about ITAR, visit: https://www.pmddtc.state.gov/

This page was intentionally left blank.

APPENDIX D

INDEPENDENT ANALYSIS OF SPACE INDUSTRIAL BASE

An independent assessment of the U.S. Space Industrial Base was performed by Quilty Analytics in July 2021. It was prepared at the request of DIU and specifically intended to provide SSIB'21 workshop attendees with a summary of emerging commercial space activities so that government and military planners can focus U.S. space priorities in ways that support our nation's strategic and commercial interests.

Quilty Analytics' full report can be downloaded here.[313]

EXECUTIVE SUMMARY

- **Rocket Science Rut** - Stubbornly high launch costs, largely unchanged for over 50 years, have been the primary barrier to the expansion of the space industry.

- **Breaking Barriers** - Over the past decade, launch costs have been lowered by an order of magnitude, thus laying the foundation for the emergence of a new, expansive space economy.

- **Role Reversal** - With this transition, we expect commercial economic activity to overtake and, in the long-term, greatly exceed government spending on space activities.

- **Earth Focused** - Today, the bulk of commercial space activities are oriented toward delivering services (primarily communications and Earth observation) to customers on the Earth.

- **Slipping the Surly Bonds** - Enabled by lower launch costs, new and yet undetermined, industries will emerge in LEO and Cislunar space.

- **Building a New Space Economy** - These emerging industries will deliver new products and services to Earth (i.e., pharmaceuticals, space-based solar power, etc.) but will also support self-sustaining economic activities within Cislunar space and the Moon.

- **Laying a New Foundation** - Today's underlying space infrastructure is insufficient to support the development and growth of these new, emerging activities and industries. The U.S. Government, working in conjunction with private industry, should focus its investment efforts on building the space infrastructure needed to support the future space economy.

- **Riding the Wave** - Recognizing the paradigm shift underway, the U.S. Government must decouple from traditional procurement and regulatory practices and embrace new engagement models that leverage commercial sector innovation and investment to outmaneuver and outpace adversaries.

---

[313] https://www.quiltyanalytics.com/wp-content/uploads/210714_QUILTY_Emerging-Space-Economy.pdf

GENERAL OBSERVATIONS

Key attributes of the space industry that set it apart from most other terrestrial industries include:

- **Protracted Development Cycles** - Flight heritage, which can only be achieved in the vacuum of space, does not come often or cheaply. Unlike terrestrial products that can be cycled through labs in days, gaining access to space has traditionally been measured in months or years.

- **Open-Loop Testing** - Prior to SpaceX's successful recovery of a first-stage booster in 2015, the "absence of failure" was the primary means of measuring a rocket's design sufficiency. All other space hardware suffers the same constraint due to a limited ability to observe, manipulate, repair, and test hardware once launched to space.

- **High Capital Intensity** - Space hardware requires careful and specialized design techniques to survive the ride to space, as well as the harsh thermal and radiation environment once there. A trip to space traditionally costs $10,000 to $20,000 per kilogram.

HOW AND WHERE SHOULD GOVERNMENT INVEST?

- **JUST BUY STUFF! And Buy Off-the Shelf** - Avoid the temptation to create soul-sucking programs of record. Become a buyer of products and services. Wherever possible, select off-the-shelf products and solutions that meet 80% or more of mission requirements instead of spending exorbitant amounts for 100% effective solutions.

- **Spread the Wealth, Move Fast** - Engage with a broad cross-section of small-to-mid-sized innovative companies using acquisition approaches that minimizes paperwork and knowledge of government procurement systems. Emphasize annual technology "sprints" that emphasize government technology priorities, emphasize a rapid tech refresh cycle, and provide an opportunity to gain heritage through "free rides" into space.

- **Buy Stuff from Allies and Partners** - While buying American should be the top priority, don't overlook the potential contribution of friends and allies that both need U.S. security support, and are capable contributors in the space domain. This concept is not inherently in conflict with "Buy American" priorities in cases where suitable technologies are not available domestically. In fact, sending a buying signal to the market will induce further domestic investment in relevant technologies and capabilities.

Collectively, these recommendations are intended to convey a simple but important message. To remain relevant in the rapidly transforming space industry, the government must move away from its traditional acquisition approach and adopt a more venture-like attitude toward investment and acquisition. This transition could entail suffering some outright failures to achieve one spectacular home run.

APPENDIX E
HYBRID SPACE ARCHITECTURE[314]

Statement of Principles

U.S. Government and commercial space capabilities are vital to our national and economic security. They are increasingly threatened militarily by potential adversaries and commercially by foreign government backed competition. To meet these challenges and retain U.S. primacy in space, the U.S. Government should partner with the U.S. space industry to rapidly transition to a Hybrid Space Architecture.

The Hybrid Space Architecture is the integration of emergent "new space" smallsat capabilities with traditional U.S. Government space systems.

This evolving resilient architecture will use a "variable trust" network framework for rapid and secure data exchange among proliferated satellite systems and services that are large and small; government and commercial; U.S. and Allied; in various, diverse, and layered orbits. The architecture shifts from a platform-centric to an information-centric paradigm.

The Hybrid Space Architecture will dramatically improve deterrence and resilience in space while providing substantial new information advantage for science, commerce, and security.

- **Distribute Risk** - provides strength in numbers and diversity, mitigating the inherent vulnerability associated with small numbers of high value assets in the current architecture.
- **Operate and Innovate Faster** – allows for more rapid collection and dissemination of vital information, as well as rapid insertion of new technologies as they mature. For the military, it allows some missions currently performed by aircraft to be hosted in space.
- **Improve Interoperability** – improves decentralized interoperability among U.S. and Allied military services; the intelligence community; civil and commercial space.
- **Lead the New Space Economy** - strengthens the U.S. commercial space economy, further boosting U.S. space leadership.

The Hybrid Space Architecture will leverage:

- Multi-path, adaptative, secure communications; open mission systems; common standards
- Edge processing; autonomous command and control/tip and cue; artificial intelligence;
- Distributed ledgers (e.g. blockchain)
- Terrestrial and space-based cloud infrastructure and analytics
- Commercial space manufacturing efficiencies (e.g. additive manufacturing and scale), systems, and data; digital modeling, design, and engineering; standards for cyber protection and secure supply chains; Agile/DevOps software and hardware approaches
- Low cost commercial bulk launch; responsive and resilient small launch
- New rapid government acquisition mechanisms to move quickly to the new architecture

---

[314] Source: https://smallsatalliance.org/

This page was intentionally left blank.

APPENDIX F
ACRONYMS & ABBREVIATIONS

3D - Three Dimensional (printing)
5G - Fifth Generation Wireless Internet
AFRL – Air Force Research Lab
AFRL/RV – Air Force Research Lab Space Vehicles Directorate
AFPC – American Foreign Policy Council (think thank)
AFSPC – Air Force Space Command (now USSF)
AI – Artificial Intelligence
AI/ML – Artificial Intelligence/Machine Learning
AOR – Area of Responsibility (DOD)
ASAT – Anti Satellite
ASDRE – Associate Director for Research and Engineering [DOD]
AST – Office of Commercial Space Transportation [FAA]
AUM – Assets Under Management
BEA – Bureau of Economic Analysis
CARES – Coronavirus Aid, Relief, and Economic Security Act
CASC – China Aerospace Science and Technology Corporation
CCL – Commerce Control List [DOC]
CFTC – Commodity Futures Trading Commission
CISA – Cybersecurity and Infrastructure Security Agency
CLPS – Commercial Lunar Payload Services (NASA)
COA – Course of Action
COCOM – Combatant Command (DOD)
CONOPS – Concept of Operations (DOD)
COTS –Commercial Orbital Transportation System (NASA)
COVID-19 – Coronavirus Disease 2019
CPC – Communist Party of Chica (CCP)
CSCO – Commercial Satellite Communications Office (USSF)
CSPO –Commercial Systems Program Office (NRO)
CSIS – Center for Strategic and International Studies (think tank)
CSPC – Center for the Study of the President and Congress (think tank)
CSRO – Chief of Space Operations Staff [USSF]
DARPA – Defense Advanced Research Projects Agency [DOD]
DevSecOps – development, security, and operations
DFC – U.S. International Development Finance Corporation
DHS – Department of Homeland Security
DISA – Defense Information Systems Agency [DOD]
DIU – Defense Innovation Unit [DOD]
DLA – Defense Logistics Agency
DOC – Department of Commerce
DOD – Department of Defense
DOE – Department of Energy
DOEd – Department of Education
DOS – Department of State
DOT – Department of Transportation
DPC – Domestic Policy Council [EOP]
DRACO – Demonstration Rocket for Agile Cislunar Operations (DARPA)
DSN – Deep Space Network
EAR – Export Administration Regulations
EO/IR – Electro Optical / Infrared (camera)
EOP – Executive Office of the President
F2T2EA – Find, Fix, Target, Track, Engage, Assess
FAA – Federal Aviation Administration [DOT]
FBI – Federal Bureau of Investigation
FCC – Federal Communication Commission
FFRDC – Federally Funded Research and Development Center
FOCI – Foreign Ownership, Control or Influence
FY – Fiscal Year
G7 – Group of Seven (states)
GAO – General Accounting Office

GEO – Geostationary Earth Orbit
GEOINT – Geospatial Intelligence
GPS – Global Positioning System
HLS – Human Landing System (NASA)
HSA – Hybrid Space Architecture
IAA – International Academy of Astronautics
IAA – Incident Awareness and Assessment
IC – Intelligence Community
ICAO – International Civil Aviation Organization
IMF – International Monetary Fund
IRL – Integration Readiness Level
IRL – Investment Readiness Level
ISR – Intelligence, Surveillance & Reconnaissance
ISS – International Space Station
ITAR – International Trafficking in Arms Regulation
JADC2 – Joint All Domain Command and Control
JROC – Joint Requirements Oversight Council
L2 – Lagrange Point Two
LEO – Low Earth Orbit
MEO – Middle Earth Orbit
MEV – Mission Extension Vehicle
MOU – Memorandum of Understanding
MRL – Manufacturing Readiness Level
MT – Metric Tons
NASA – National Aeronautics and Space Agency
NATO – North Atlantic Treaty Organization
NDAA – National Defense Authorization Act
NDRC – National Development and Reform Commission [China]
NEC – National Economic Council [EOP]
NEO – Noncombatant Evacuation
NGA – National Geospatial Agency [DOD]
NIAC – NASA Innovative Advanced Concepts
NIST – National Institute of Standards and Technology
NMS – National Military Strategy
NOAA – National Oceanic and Atmospheric Agency [DOC]
NRL – Naval Research Lab [DOD]
NRO – National Reconnaissance Organization [DOD]

NSC – National Security Council [EOP]
NSF – National Science Foundation
NSIC - National Security Investment Capital [DOD]
NSpC – National Space Council [EOP]
NSS – National Security Strategy
NSSL – National Security Space Launch
NTIA –National Telecommunications and Information Administration
OECD – Organization for Economic Cooperation and Development
ODNI – Office of the Director of National Intelligence
OMB – Office of Management and Budget [EOP]
OPR – Office of Primary Responsibility
OSAM – On-Orbit Servicing Assembly and Manufacturing
OSC – Office of Commercial Space [DOC]
OSD – Office of the Secretary of Defense [DOD]
OTMP– Office of Trade and Manufacturing Policy [EOP]
OSTP – Office of Science and Technology Policy [EOP]
OTA – Other Transaction Authority
OUSD R&E – Office of the Undersecretary of Defense for Research and Engineering
PNT – Positioning, Navigation and Timing
POTUS – President of the United States
PPP – Paycheck Protection Program
PR – Personnel Recovery
PRC – People's Republic of China
QSB – Qualified Small Business
R&D – Research and Development
RDT&E – Research Development Test and Evaluation
RF – Radio Frequency
RLSP – Rocket Systems Launch Program
ROS – Robot Operating System
ROTC – Reserve Officer Training Corps
RPO – Rendezvous and Proximity Operations
RSGS – Robotic Servicing of Geosynchronous Satellites [DARPA]
S&T – Science and Technology
SAR – Synthetic Aperture Radar
SATCOM – Satellite Communications
SBA – Small Business Administration

SBIR – Small Business Innovative Research
SDA – Space Development Agency
SDA – Space Domain Awareness
SEC – Securities & Exchange Commission
S-ISAC – Space Information Sharing and Analysis Center (IC)
SIS – Space Information Services
SMC – Space and Missile Systems Center [USSF] now SSC
SPAC – Special Purpose Acquisition Corporation
SpaceX – Space Exploration Technologies (company)
SpEC – Space Enterprise Consortium
SPO – System Program Office
SPR – Strategic Petroleum Reserve
SSA – Space Situational Awareness
SSC – Space Systems Command [USSF] formerly SMC
SSP – Space Solar Power
SSIB – State of the Space Industrial Base (report)
SSPIDR – Space Solar Power Incremental Demonstrations and Research Project [AFRL]

STEM – Science Technology Engineering and Math
STTR – Small Business Technology Transfer (STTR)
TLE – Two-Line Element (Astronautics)
TLI – Trans-Lunar Injection
TRL – Technology Readiness Level
UAE – United Arab Emirates
UK – United Kingdom
ULA – United Launch Alliance (company)
U.S. – United States
USG – United States Government
USGS – United States Geological Service
USML – U.S. Munitions List [DOS]
USSF – United States Space Force [DOD]
USSPACECOM – United States Space Command [DOD]
VC – Venture Capital(ist)
VP – Vice President of the United States
WMO – World Meteorological Organization
WWII – World War Two
XGEO – Beyond Geostationary Orbit

U.S. Department *of* Defense

Printed in the USA
CPSIA information can be obtained
at www.ICGtesting.com
LVHW071811240124
769490LV00077B/2557